Metaverse Platform Implementation

This book provides a comprehensive exploration of the engineering and scientific principles behind the Metaverse, a 3D digital world designed to replicate and enhance physical reality through some advanced technologies, such as extended reality (xR), artificial intelligence (AI), blockchain, 5G/6G, and computer vision. The book delves into the hardware (sensors, headsets, edge/cloud servers) and software (3D scene reconstruction, synchronization, AI processing) components that form the backbone of Metaverse platforms, addressing critical aspects such as security, privacy, and social implications.

This book is organized into three parts and covers some foundational technologies, such as neural rendering and semantic communications, security challenges (involving attack models and countermeasures), and real-world applications in healthcare (such as mixed reality-based stroke rehabilitation and telehealth innovations). By linking virtual and augmented reality to the broader Metaverse ecosystem, this book serves as a crucial resource for engineers, scientists, and researchers in academia or industries working in the fields of robotics, smart cities, and intelligent transportation, offering valuable insights into the development of secure, efficient, and immersive digital 3D internet environments.

Metaverse Platform Implementation

AI, Security, and Applications

Edited by
Fei Hu, Shengting Cao, and Xiali (Sharon) Hei

CRC Press
Taylor & Francis Group
Boca Raton London New York

CRC Press is an imprint of the
Taylor & Francis Group, an **informa** business

Designed cover image: Shutterstock Image ID 2062865573

First edition published 2026
by CRC Press
2385 NW Executive Center Drive, Suite 320, Boca Raton FL 33431

and by CRC Press
4 Park Square, Milton Park, Abingdon, Oxon, OX14 4RN
CRC Press is an imprint of Taylor & Francis Group, LLC

ISBN: 978-1-032-66355-5 (hbk)
ISBN: 978-1-032-66358-6 (pbk)
ISBN: 978-1-032-66360-9 (ebk)

DOI: 10.1201/9781032663609

Typeset in Sabon
by Deanta Global Publishing Services, Chennai, India

Contents

PART II
Security Issues in the Metaverse

PART III
Metaverse Applications

Preface

Since the invention of the internet more than four decades ago, numerous engineers/scientists have been thinking of the possibility of implementing a 3D internet, that is, the Metaverse, to allow people to interact with a digital world that is exactly the same as our physical world. The Metaverse relies on the success of many advanced cyberspace technologies, such as extended reality (xR), artificial intelligence (AI), blockchain, 5G/6G, cybersecurity, Internet of Sensors, and computer vision.

Unlike existing books on the Metaverse, this book focuses on the detailed engineering/science implementation of Metaverse platforms, including its hardware assembly (sensors, head-mounted displays, cameras, edge/cloud servers, etc.) and software modules (3D scene capture/reconstruction, multi-site synchronization, task scheduling, spatial referencing, data communications, AI processing, blockchain, etc.). It also covers security/privacy, social acceptability, ethics, and Metaverse applications (healthcare, education, industry, transportation, etc.). This book clearly describes the evolution from virtual reality, augmented reality, to the Metaverse. It explains the Metaverse ecosystem with avatar design, multiuser content creation, virtual trading, privacy threats, and trust issues. It emphasizes the integration of edge computing, robotics, AI, and computer vision. The applications use the Metaverse to implement telepresence, convenient 3D construction, seamless social reactions, education during pandemics, etc.

The book underscores the following four aspects: (1) It emphasizes the engineering/science design through various models/algorithms. (2) It describes the seamless integration of xR, AI, sensing, computer vision, and networking. (3) It highlights security and privacy issues in Metaverse platforms. (4) It presents interesting Metaverse applications with a detailed testing methodology.

The book consists of three parts:

PART I: BASIC TECHNOLOGIES FOR THE METAVERSE

Part I comprises six chapters on the fundamental technologies underpinning Metaverse platforms. For example, Chapter 1 explains all necessary hardware units for a basic VR communication system. It particularly elucidates neural rendering technology, a groundbreaking advancement in image synthesis and 3D reconstruction, leveraging the power of artificial neural networks to manipulate or generate images, videos, and 3D models.

Chapter 2 details the technique of reconstructing the 3D geometry and structure of an object from one or multiple images. The primary challenge lies in the ambiguity arising from the absence of depth and semantic information. For human body reconstruction, a robust reconstruction technique should fulfill several criteria, including high resolution, efficiency, accuracy, and adaptability to various poses and clothing styles on different bodies.

Chapter 3 introduces a unified framework for 3D point cloud segmentation. By integrating transformer-based models such as KPConv, SphereFormer, PReFormer, and Cylinder3D, the approach achieves high accuracy and efficiency across structured and unstructured scenes. Experiments on SemanticKITTI, RELLIS-3D, and Toronto3D demonstrate strong performance with reduced computational costs, enabling real-time scene understanding for immersive virtual environments.

Chapter 4 examines how wireless communication and AI-driven techniques help improve the Metaverse, focusing on how semantic communication and AI enhance user interactions, foster deeper comprehension, and drive immersive engagement. We delve into various aspects, including semantic technologies, UAV applications, natural language processing, virtual embodiment, and non-orthogonal multiple access (NOMA), highlighting their integration to elevate user engagement and collaborative potential.

Chapter 5 explores the communication technologies essential for enabling the Metaverse, focusing on emerging high-bandwidth, low-latency solutions. It highlights ongoing research in protocols, hardware, and software development, emphasizing the need for robust communication infrastructure. By examining a range of promising technologies across diverse paradigms, the chapter outlines how these innovations form the foundation for a fully realized Metaverse.

Chapter 6 explores the emotional and ethical dynamics of human interactions with 3D avatars in digital environments, incorporating a range of moral, religious, and philosophical perspectives. We investigate the development of deep emotional bonds with virtual humans and analyze how these relationships interact with traditional values and various religious doctrines.

PART II: SECURITY ISSUES IN THE METAVERSE

Part II examines comprehensive models and solutions addressing numerous threats on Metaverse platforms. Chapter 7 discusses a fundamental issue concerning Metaverse security and privacy, as well as multiple attack scenarios and solutions for general and specific attacks. We then survey the major challenges in developing the Metaverse, including potential attack scenarios and vulnerable sections due to inadequate security measures. By discussing several possible Metaverse attacks, as well as potential solutions, this chapter provides insightful insights and valuable guidelines for researchers and Metaverse developers.

Chapter 8 examines the complex security challenges of the Metaverse, outlining strategies across privacy protection, machine learning safeguards, and decentralized cryptographic solutions. It highlights emerging threats and countermeasures—from telemetry-based attacks to blockchain-enabled trust mechanisms—emphasizing the need for ethical, user-centric, and scalable security designs to ensure safe and trustworthy immersive experiences.

Chapter 9 further explains various attack models in the Metaverse. It explores the motivation behind enhancing security measures for Metaverse platforms, identifies three prominent attack models targeting its hardware and software units, and discusses contemporary countermeasures to address these threats. Additionally, the chapter outlines future research trends in Metaverse security and highlights key challenges that must be overcome to establish a secure and resilient Metaverse environment.

PART III: METAVERSE APPLICATIONS

Part III provides some interesting application scenarios for Metaverse platforms.

Chapter 10 provides a comprehensive overview of the Metaverse architecture with a focus on AR/VR environments, highlighting key security and privacy threats. It explores adversarial attacks and mitigation strategies using AI, behavioral security, and blockchain technologies. The chapter also outlines open challenges and future research directions, offering a foundational understanding of the Metaverse's components and associated risks.

Chapter 11 introduces one of our projects on telehealth. This project utilizes mixed reality to remotely simulate in-person doctor's appointments with a focus on stroke rehabilitation. Doctors employ the system to evaluate the physical and cognitive skills of post-stroke patients. Patients complete tasks within several mixed reality video games that utilize the KineAssist treadmill as a controller deployed to the Microsoft HoloLens 2.

Chapters 12 and 13 provide insight into the application of Metaverse for telerehabilitation from different perspectives. Chapter 14 discusses the digital twin approach and the cybersecurity considerations of the rehabilitation applications. Some technological challenges are explained and the corresponding solutions are provided.

It is noteworthy that this will be the first book on the Metaverse system, focusing on system hardware/software components, security and privacy solutions, and its applications. It has many valuable ideas for engineering/science design in Metaverse. It links VR, AR, and Metaverse seamlessly by using some practical telehealth applications.

Targeted Audiences: Currently, there are more than 20 million people in the world involved in the R&D process of Metaverse products. This book will attract the attention of engineers and scientists in the industry of vehicles, robots, smart cities, intelligent transportation, etc. It will also be of interest to millions of academicians (faculty/students, researchers) working on Metaverse systems.

Editor Biographies

Fei Hu has authored over ten technical publications with CRC Press. His research focus includes cyber security and networking. He is currently a professor in the Department of Electrical and Computer Engineering at the University of Alabama (UA), Tuscaloosa, Alabama, USA. He earned his Ph.D. in the field of signal processing (in 1999) from Tongji University (Shanghai, China) and in electrical and computer engineering (ECE, in 2002) from Clarkson University (New York, USA). He has published over 200 journal/conference papers and books. Dr. Hu's research has been supported by the U.S. National Science Foundation, Cisco, Sprint, and other sources.

Shengting Cao is a Ph.D. candidate in the ECE department at The University of Alabama and will graduate in 2025. His research focus is the Metaverse. He has published several IEEE transaction papers to date. He has developed a set of 3D object reconstruction algorithms and tools. He has invented the technology that utilizes a single-belt treadmill as a substitute for dual-belt treadmill during rehabilitation training.

Xiali (Sharon) Hei has been an Alfred and Helen M. Lamson Endowed Associate Professor in the School of Computing and Informatics at the University of Louisiana at Lafayette since August 15, 2023. She was an Alfred and Helen M. Lamson Endowed Assistant Professor in the School of Computing and Informatics at the University of Louisiana at Lafayette from August 2017 to August 2023. During her sabbatical year of 2023–2024, she visited Dr. Insup Lee at the University of Pennsylvania. Prior to joining the University of Louisiana at Lafayette, she was an assistant professor at Delaware State University from 2015 to 2017 and an assistant professor at Frostburg State University from 2014 to 2015. Sharon received her Ph.D. in computer science from Temple University in 2014, focusing on computer security.

Contributors

Timothy Adewale Adeyi
Nanjing University of Information Science and Technology
Nanjing, China

Nashmi Alheraijy
University of Alabama
Tuscaloosa, USA

Qais Altannak
University of Alabama
Tuscaloosa, USA

Trent Boughner
University of Alabama
Tuscaloosa, USA

Patrick Brooks
University of Alabama
Tuscaloosa, USA

Luke Burbee
University of Alabama
Tuscaloosa, USA

Jun Cai
Nanjing University of Information Science and Technology
Nanjing, China

Aidan Cantu
University of Alabama
Tuscaloosa, USA

Shengting Cao
University of Alabama
Tuscaloosa, USA

Adrian David Cheok
Nanjing University of Information Science and Technology
Nanjing, China

Jingyang Cui
University of Alabama
Tuscaloosa, USA

Fei Hu
University of Alabama
Tuscaloosa, USA

Md Rafi Islam
University of Alabama
Tuscaloosa, USA

Md Imtiaz Kamrul
University of Alabama
Tuscaloosa, USA

Shuhui Li
University of Alabama
Tuscaloosa, USA

Weiqiang Lyu
University of Alabama
Tuscaloosa, USA

Md Nurunnabi
University of Alabama
Tuscaloosa, USA

Zane Pautzke
University of Alabama
Tuscaloosa, USA

Shehan Irteza Pranto
University of Alabama
Tuscaloosa, USA

Kenneth Ricks
University of Alabama
Tuscaloosa, USA

Shunqiao Sun
University of Alabama
Tuscaloosa, USA

Xue Wu
University of Alabama
Tuscaloosa, USA

Lifan Xu
University of Alabama
Tuscaloosa, USA

Ying Yan
Nanjing University of Information Science and Technology
Nanjing, China

Yongzhi Yang
University of Alabama
Tuscaloosa, USA

Ruxin Zheng
University of Alabama
Tuscaloosa, USA

Part I

Basic Technologies for the Metaverse

Chapter 1

Neural Rendering Technology in VR and AR

Shengting Cao and Xue Wu

1.1 INTRODUCTION

1.1.1 Background

Virtual reality (VR) and augmented reality (AR) have experienced significant advancements over the past few decades. What began as experimental technology in the mid-twentieth century has now evolved into immersive platforms that blend physical and digital worlds. This transformation has been driven by key developments in hardware and computational techniques, which have enhanced the realism, interactivity, and accessibility of VR/AR systems.

Figure 1.1 illustrates the evolution of VR and AR hardware [1]. The introduction of Morton Heilig's *Sensorama* [2] in the 1960s provided a glimpse into the possibilities of multisensory immersion, integrating 3D visuals with sound and motion. The journey of wearable VR hardware began with Ivan Sutherland's "Sword of Damocles" in the 1960s – a head-mounted display (HMD) that could render rudimentary 3D graphics [3]. However, this early system was limited by its weight and mechanical tracking, making it impractical for widespread use. In the following decades, the focus shifted toward creating more lightweight, mobile, and responsive systems.

The 1990s marked a significant leap with the creation of projection-based systems such as the Cave Automatic Virtual Environment (CAVE) [4]. Unlike earlier head-mounted systems, the CAVE used room-scale projections to create semi-immersive virtual environments. However, consumer-grade VR hardware remained out of reach until the early 2010s, with the release of devices such as the Oculus Rift DK1 [5]. The Oculus Rift combined high-resolution displays with gyroscopic sensors, providing a relatively affordable and accessible VR experience.

Following the success of the Oculus Rift [5], devices such as the HTC VIVE [6] and PlayStation VR [22] introduced room-scale tracking and advanced input controllers, further refining the VR experience. In parallel, Microsoft's HoloLens (2016) [7] and Magic Leap's [8] AR headsets

DOI: 10.1201/9781032663609-2

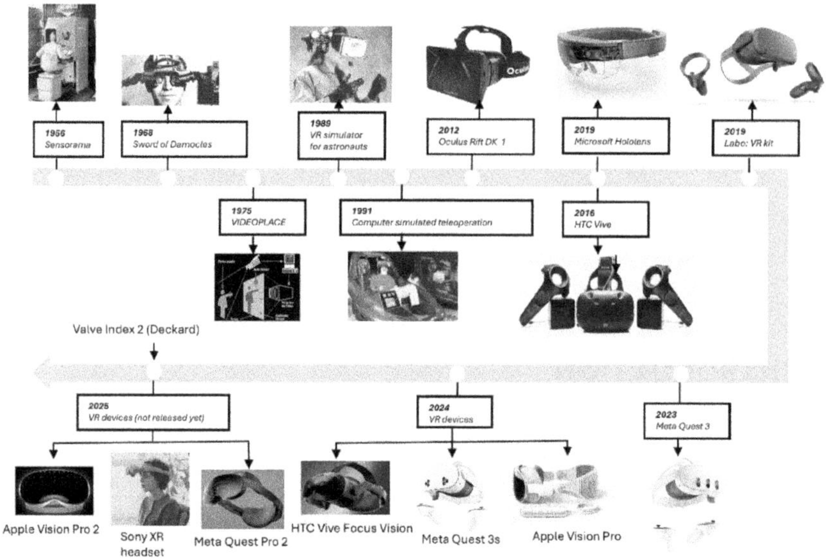

Figure 1.1 Evolution of VR and AR hardware over time.

advanced the AR domain by integrating real-time spatial mapping and holographic projection into consumer products.

In recent years, modern systems such as Meta Quest 3 [9] and Apple Vision Pro [10] have pushed the boundaries of immersive technology. These devices offer standalone functionality, eliminating the need for external personal computer while integrating powerful system-on-chip (SoC) architectures with dedicated neural processing units (NPUs). These NPUs support real-time scene reconstruction and neural rendering [11] by handling large-scale neural network computations locally, enhancing performance without the constraints of wired connections.

1.1.2 Objectives of the Chapter

This chapter aims to address the role of neural rendering in creating realistic, immersive experiences and emphasize the importance of hardware evolution in supporting this emerging technology.

1.2 EVOLUTION OF VR/AR HARDWARE AND ITS ROLE IN NEURAL RENDERING

The evolution of VR and AR hardware has been driven by the pursuit of more immersive, realistic, and seamless user experiences. Key hardware innovations have enhanced display quality, frame rates, spatial tracking, and computational power, all of which are critical to the success of neural rendering systems. As neural rendering shifts the paradigm of how scenes are rendered in real time, it relies on robust hardware to handle the computational demands of deep learning models that generate high-fidelity 3D visuals from sparse input data. This section outlines the milestones in VR and AR hardware that have contributed to the growth of neural rendering capabilities.

The advancements in display resolution, frame rates, tracking precision, and computational capabilities have played a pivotal role in shaping modern VR/AR systems and enabling neural rendering.

1.2.1 Display Resolution and Frame Rates

Early VR systems, such as those in the 1990s, featured low-resolution displays that limited visual fidelity and immersion. Modern headsets, such as the HTC VIVE Pro, Meta Quest 3, and Apple Vision Pro, now offer resolutions exceeding 4K per eye, dramatically improving visual clarity. These higher resolutions are crucial for neural rendering frameworks, which produce detailed textures, complex lighting, and photorealistic scenes. Frame rate improvements have also been significant; earlier systems operated at 30–60 Hz, while today's devices achieve refresh rates of 90–120 Hz. High frame rates are essential for preventing motion sickness and ensuring fluid, immersive interactions.

1.2.2 Spatial Tracking Precision

Accurate spatial tracking is a cornerstone of effective VR and AR systems. Early HMDs relied on external infrared trackers, while newer headsets use inside-out tracking with embedded cameras and sensors. Systems such as the HTC VIVE introduced precise room-scale tracking with lighthouse sensors, whereas HoloLens and Meta Quest 3 integrated real-time depth mapping to localize the user within their environment. This improvement in spatial tracking supports neural rendering by enabling real-time scene updates that align perfectly with the user's movements, creating seamless interactions.

1.2.3 Computational Power

Processing power has grown exponentially with the integration of advanced SoC architectures and NPUs [12]. Modern headsets such as the Meta Quest 3 feature SoCs with dedicated cores for handling machine learning tasks, making real-time neural rendering feasible. Graphics processing units (GPUs) in devices such as the Apple Vision Pro support teraflops of

Figure 1.2 HTC VIVE.

operations per second, enabling neural networks to reconstruct complex 3D scenes with minimal latency. These computational advancements ensure that neural rendering models such as Neural Radiance Fields (NeRF)[11] can operate smoothly without offloading tasks to external servers.

1.3 HTC VIVE: A MILESTONE IN VR HARDWARE

The HTC VIVE as shown in Figure 1.2, released in 2016, was a landmark in the history of consumer VR, setting a new standard for immersion and precision. The system featured high-resolution OLED displays with a 90 Hz refresh rate and room-scale tracking, which allowed users to move freely within a designated physical space while interacting with virtual environments.

One of the defining features of the HTC VIVE was its external "lighthouse" tracking system. This system used two base stations that emitted infrared signals to detect the position and orientation of the headset and controllers with sub-millimeter precision. This level of tracking accuracy was unprecedented and provided a breakthrough in spatial localization, laying the groundwork for neural rendering applications that depend on precise positional data to align virtual and physical elements in real time.

In addition to precise spatial tracking, the VIVE's controllers featured motion sensors that enabled users to interact with virtual objects naturally. This interactivity is crucial for neural rendering systems, which often require user input to navigate dynamic scenes or adjust viewpoints. The HTC VIVE demonstrated that real-time, interactive environments could be achieved on consumer hardware, paving the way for subsequent headsets to incorporate neural rendering capabilities.

1.4 HOLOLENS: KEY AR HARDWARE COMPONENTS

Microsoft's HoloLens, first introduced in 2016 [13], revolutionized AR by combining wearable AR technology with real-time spatial mapping. Unlike VR headsets, which immerse users in fully virtual worlds, HoloLens overlays digital content onto the real world using transparent displays.

As shown in Figure 1.3, the HoloLens' key hardware components include a depth-sensing camera, an inertial measurement unit (IMU), and multiple spatial microphones. The depth-sensing camera uses structured light to map the physical environment in three dimensions, creating a spatial mesh that neural rendering systems can use for object placement and collision detection. This real-time mapping is essential for AR applications that require virtual objects to interact believably with real-world elements.

Another notable feature of the HoloLens is its holographic display system, which supports dynamic adjustments in brightness and contrast to maintain visual consistency regardless of ambient lighting conditions. This capability enhances neural rendering outputs by ensuring that virtual objects blend seamlessly with their real-world surroundings, maintaining the illusion of coexistence.

Additionally, HoloLens incorporates a built-in NPU that processes spatial data and supports real-time scene reconstruction. This architecture enables neural rendering frameworks to operate locally on the device, reducing latency, and ensuring that the user's interactions are synchronized with the rendered visuals.

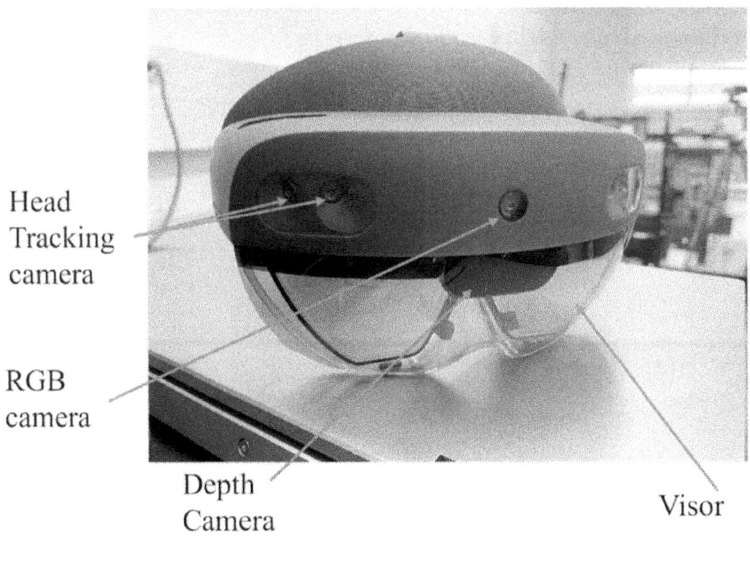

Figure 1.3 HoloLens 2.

1.5 META QUEST 3: BRIDGING STANDALONE VR AND NEURAL RENDERING

As shown in Figure 1.4, the Meta Quest 3, launched in 2023, represents a significant leap in standalone VR headset technology. Unlike its predecessors, the Quest 3 is equipped with an advanced Snapdragon XR2 Gen 2 chip that integrates Central Processing Unit (CPU), GPU, and NPU components for efficient handling of complex rendering tasks. This hardware configuration enables Quest 3 to support real-time neural rendering without the need for external processing power.

One of the standout features of the Meta Quest 3 is its high-resolution pancake lenses, which offer improved clarity and reduced lens distortion compared to traditional Fresnel lenses. This enhancement is particularly important for neural rendering, as it ensures that the system's photorealistic outputs are displayed accurately without visual aberrations.

In addition to its upgraded display, the Quest 3 features six degrees of freedom (6DoF) inside-out tracking with multiple embedded cameras. These cameras not only track the user's movements but also perform simultaneous localization and mapping (SLAM), enabling real-time updates to the virtual environment. This tracking precision supports neural rendering algorithms by providing accurate data for scene reconstruction and interaction mapping.

The Quest 3 also incorporates foveated rendering, which uses eye-tracking to prioritize rendering detail in the user's focal area while reducing detail in the peripheral regions. By optimizing computational resources in this way, Quest 3 can achieve higher frame rates and more stable performance, making it an ideal platform for neural rendering applications that demand both visual fidelity and interactivity.

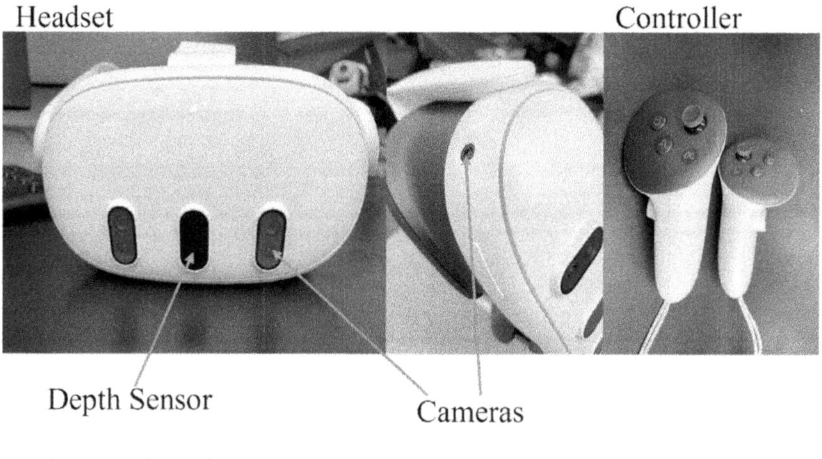

Figure 1.4 Meta Quest 3.

1.6 APPLE VISION PRO: PIONEERING SPATIAL COMPUTING

As shown in Figure 1.5, the Apple Vision Pro, introduced in 2024, represents a new era of spatial computing by integrating cutting-edge hardware designed specifically for mixed reality experiences. The Vision Pro features dual micro-OLED displays with a combined resolution of over 23 million pixels, offering unparalleled visual clarity. These displays support refresh rates of up to 120 Hz, ensuring smooth, lifelike interactions.

One of the Vision Pro's defining features is its advanced eye-tracking system, which enables precision foveated rendering. By detecting the exact point of the user's gaze, the system renders the area in focus at maximum resolution while downscaling the surrounding areas. This approach significantly reduces the computational load, allowing the device to allocate resources efficiently for neural rendering tasks.

The Vision Pro also incorporates a powerful M2 chip alongside a dedicated R1 coprocessor, which handles sensor fusion from the device's 12 cameras, 5 sensors, and 6 microphones. This architecture supports real-time neural network inference, enabling the Vision Pro to process complex scene data and render photorealistic virtual elements seamlessly into the user's environment. The combination of LiDAR and depth sensing further enhances the device's ability to create accurate spatial maps, making it an ideal platform for neural rendering in AR applications.

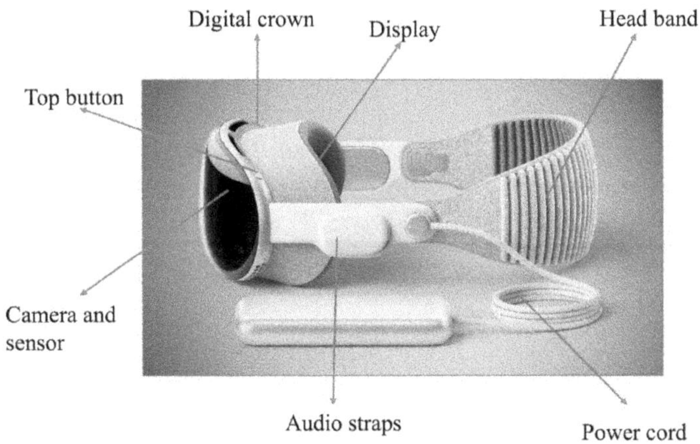

Figure 1.5 Apple Vision Pro.

1.7 TRADITIONAL RENDERING TECHNIQUES IN VR/AR

Traditional rendering techniques have played a crucial role in the development of VR/AR systems, providing a foundation for generating visual content that simulates real-world environments. These techniques rely on mathematical models and algorithms to project 3D objects onto 2D screens, creating the illusion of depth and realism. However, achieving real-time rendering at high frame rates with photorealistic detail has always been a challenging task due to hardware limitations, computational costs, and the complexity of dynamic scenes. This section provides an overview of traditional rendering methods, examines real-time rendering optimizations, and discusses the inherent limitations of these approaches in VR/AR applications.

1.7.1 Overview of Traditional Rendering

1.7.1.1 Polygonal-Based Rendering

The most widely used traditional rendering approach is polygonal-based rendering, which represents 3D objects as collections of polygons, typically triangles. This representation provides a balance between rendering speed and visual accuracy, as triangles can be processed efficiently by modern GPUs. Polygonal-based rendering has two primary techniques: rasterization and ray tracing.

1.7.1.2 Rasterization

Rasterization is the most common technique for real-time rendering, used extensively in gaming and VR/AR applications. It works by converting 3D models into 2D representations that can be displayed on a screen. The process involves projecting 3D vertices onto the screen space, creating polygons that are then filled with color, texture, and shading. The GPU processes each polygon independently, applying lighting calculations based on predefined models such as Phong shading or Gouraud shading [14]. Rasterization is fast because it processes each pixel without simulating the entire path of light rays, making it ideal for real-time applications where maintaining high frame rates is critical.

1.7.1.3 Ray Tracing

Ray tracing, on the other hand, is a technique that simulates the path of light rays as they interact with surfaces in a 3D scene. By tracing the path of each ray from the camera through the scene, ray tracing can produce realistic effects such as reflections, refractions, shadows, and global illumination.

Unlike rasterization, ray tracing calculates the physical interactions of light, making it capable of generating photorealistic images. However, the computational complexity of ray tracing is significantly higher, as it requires millions of ray-path calculations per frame. This makes real-time ray tracing challenging, though recent advancements in GPU hardware (e.g., NVIDIA's RTX series with dedicated ray-tracing cores) have made hybrid approaches possible.

1.7.1.4 Math of 3D-to-2D Projection Process

At the heart of VR/AR rendering lies the task of projecting 3D coordinates onto a 2D viewing plane while preserving spatial relationships and depth perception. This transformation is achieved through three primary stages: *model transformation, view transformation*, and *projection transformation.*

1.7.1.5 Model Transformation

Model transformation converts an object's local coordinate system into the global world coordinate system. Each 3D model is defined in its *local space*, where its origin is the model's center. To place the model within the virtual environment, we apply a model matrix M, which combines translation, rotation, and scaling operations. If $v = [x, y, z, 1]^T$ represents a vertex in local coordinates, its transformation is given by:

$$v_{world} = Mv \tag{1.1}$$

Here, M can be expressed as $\boldsymbol{M} = \boldsymbol{T} \cdot \boldsymbol{R} \cdot \boldsymbol{S}$, where T is the translation matrix, R is the rotation matrix, and S is the scaling matrix.

View Transformation

Once the model is positioned in the world, it must be viewed from the perspective of a virtual camera. This is achieved through the view matrix V, which aligns the camera's coordinate system with the world coordinate system[15]. The camera's position C and orientation R define this matrix:

$$v = \begin{bmatrix} \boldsymbol{R}^T & -\boldsymbol{R}^T\boldsymbol{C} \\ 0 & 1 \end{bmatrix} \tag{1.2}$$

Applying the view matrix transforms a vertex from world space to camera space:

$$v_{view} = Vv_{world}$$

1.7.1.6 Projection Transformation

The final step is to map the camera space coordinates onto a 2D viewing plane. This is done using a projection matrix *P*, which can either be *orthographic* or *perspective*[15]. VR/AR systems predominantly use perspective projection to mimic how humans perceive depth. The perspective projection matrix is defined as

$$P = \begin{bmatrix} \frac{1}{\tan\left(\frac{\theta}{2}\right)} & 0 & 0 & 0 \\ 0 & \frac{1}{\tan\left(\frac{\theta}{2}\right)} & 0 & 0 \\ 0 & 0 & \frac{f+n}{n-f} & \frac{2fn}{n-f} \\ 0 & 0 & -1 & 0 \end{bmatrix} \tag{1.3}$$

where

- θ: Field of view angle
- n, f: Near and far clipping planes.

To convert the projected coordinates into normalized device coordinates, we perform *perspective division*, dividing by the fourth (homogeneous) coordinate:

$$x = \frac{x'}{w'}, y = \frac{y'}{w'}, z = \frac{z'}{w'} \tag{1.4}$$

1.8 REAL-TIME RENDERING APPROACHES

Achieving real-time frame rates in VR/AR applications requires rendering frames at speeds that match or exceed the display's refresh rate, typically 90–120 frames per second (FPS). To meet these demands, developers employ various optimization techniques that reduce the computational burden without compromising the overall visual quality.

1.8.1 Level of Detail Optimization

One of the most common real-time rendering optimizations is Level of Detail (LOD) [16]. LOD techniques dynamically adjust the complexity of

3D models based on the camera's distance from the object. For objects close to the user, the system renders highly detailed meshes, while for distant objects, it uses simplified versions with fewer polygons. This reduces the number of calculations needed for rendering distant geometry, freeing up GPU resources to maintain higher frame rates.

For example, in large open-world VR environments, objects such as trees, buildings, and distant landscapes are rendered with minimal detail until the user moves closer to them. This technique is particularly important for VR applications, where a consistent frame rate is necessary to avoid motion sickness and ensure a smooth experience. Neural rendering techniques, in contrast, often bypass the need for LOD by using learned representations to synthesize high-fidelity scenes at multiple resolutions without predefining complex polygonal hierarchies.

1.8.2 Baked Lighting

Lighting is one of the most computationally expensive aspects of rendering. Dynamic lighting, which requires real-time calculations for shadows, reflections, and indirect illumination, can severely impact performance. Baked lighting is an optimization technique where certain lighting effects, such as shadows and ambient light, are precomputed and stored as textures. These "baked" textures are then applied during rendering, eliminating the need for real-time lighting calculations.

Baked lighting [17] is commonly used in static scenes where the lighting conditions do not change. For example, architectural visualizations and static VR environments often use baked global illumination to simulate realistic light bounces without the computational overhead of dynamic ray-traced lighting. However, baked lighting lacks the flexibility of dynamic lighting and cannot easily accommodate changes in light sources or object positions, limiting its application in interactive VR/AR scenarios.

1.8.3 Limitations of Traditional Rendering

Despite their widespread use, traditional rendering techniques face significant limitations, particularly in the context of VR/AR applications that require real-time performance, dynamic interactions, and photorealistic visuals.

1.8.4 Memory Constraints and Computational Costs

Polygonal-based rendering relies on high-resolution 3D models and detailed textures to achieve visual realism. These assets consume a large amount of memory, especially in VR applications where stereoscopic rendering requires rendering two slightly different views for each eye. High-resolution textures, complex shaders, and detailed models can quickly exhaust the available GPU memory, leading to frame drops and degraded performance.

Ray tracing, despite its visual advantages, is even more memory-intensive, as it requires storing and processing vast amounts of geometric and lighting data. The computational cost of tracing millions of rays per frame makes real-time ray tracing infeasible for most standalone VR/AR headsets, which have limited processing power compared to desktop GPUs.

1.8.5 Trade-Offs Between Realism and Performance

Traditional rendering often requires compromises between realism and performance. To achieve the high frame rates necessary for VR, developers frequently reduce the quality of shadows, reflections, and textures, resulting in less realistic visuals. For example, rasterization-based rendering cannot easily simulate complex light interactions such as caustics or subsurface scattering, leading to visual artifacts in scenes with glass, water, or translucent materials.

While baked lighting can improve performance by precomputing lighting effects, it sacrifices the ability to interact with dynamic lighting changes, limiting its use in interactive scenes. These trade-offs are particularly noticeable in VR/AR, where the user's sense of presence is heavily influenced by the realism and responsiveness of the environment.

1.8.6 Challenges in Rendering Dynamic and Complex Scenes

Dynamic scenes involving moving objects, changing light sources, and user interactions pose [24] significant challenges for traditional rendering techniques. Real-time lighting and shadow calculations require substantial computational resources, often leading to latency issues that can disrupt the immersive experience. Additionally, rendering large-scale, complex scenes with numerous interacting elements can lead to bottlenecks, as the system must process and update each element individually.

In contrast, neural rendering approaches have shown promise in addressing these challenges by learning to approximate scene details and dynamically synthesizing realistic visuals without explicitly simulating every interaction. NeRF, for example, can generate view-consistent images from sparse input data, potentially reducing the need for complex geometric representations and traditional lighting calculations

1.9 FUNDAMENTALS OF NEURAL RENDERING

Neural rendering represents a paradigm shift in the creation of immersive digital environments, leveraging neural networks to synthesize 3D scenes and dynamic viewpoints from limited data. Unlike traditional rendering methods, which rely on polygonal meshes and complex light simulations, neural rendering encodes scenes as mathematical functions learned through deep learning. A cornerstone of this approach is NeRF, a breakthrough technology that reconstructs 3D scenes with unprecedented fidelity using sparse 2D image inputs. This section explores the fundamentals of neural rendering, including an in-depth overview of NeRF, its pipeline, and a comparison with traditional rendering techniques.

1.9.1 Introduction to Neural Radiance Fields

NeRF are a neural network-based approach for 3D scene reconstruction and novel view synthesis. Developed to address the limitations of traditional 3D modeling and rendering, NeRF encodes an entire 3D scene into a neural network, which can then render photorealistic images from arbitrary viewpoints by learning how light interacts with objects in that space. Unlike polygon-based methods, which use explicit geometric representations, NeRF uses a fully implicit representation, mapping 3D spatial coordinates and viewing directions to color and density values.

1.9.2 Role of NeRF in 3D Scene Reconstruction and View Synthesis

NeRF's primary function is to reconstruct 3D scenes by learning from a set of 2D images taken from different viewpoints. The model estimates how light radiates from different points in 3D space based on the perspective of the camera. This allows NeRF to generate highly realistic images that include subtle lighting effects, such as soft shadows, reflections, and global illumination.

One of NeRF's most groundbreaking contributions is its ability to perform novel view synthesis. Once trained, the network can render images from viewpoints that were not included in the original input dataset. This capability is crucial for VR and AR applications, where the user's position and perspective change constantly, requiring real-time updates to the scene.

1.9.3 Representing 3D Scenes through Neural Networks Trained on 2D Image Inputs

NeRF represents a 3D scene as a continuous volumetric function that maps spatial coordinates (x, y, z) and viewing directions (θ, φ) to color (R, G, B) and density values. The network is trained using a set of 2D input images, each accompanied by the corresponding camera parameters (e.g., position, orientation, and focal length). During training, the network learns to approximate the color and density at each 3D point by minimizing the difference between the rendered output and the ground truth 2D images.

By encoding 3D scenes in this way, NeRF eliminates the need for explicit geometric meshes and pre-baked lighting. Instead, the network synthesizes the appearance of objects and lighting effects directly from the data. This implicit representation is particularly advantageous for scenes with complex geometries, such as foliage, semi-transparent objects, and non-uniform lighting, which are difficult to model accurately using polygonal meshes

1.9.4 NeRF Pipeline

The NeRF rendering process can be broken down into three key stages: input image collection, neural network training, and rendering novel views. Each stage plays a critical role in enabling the system to generate photorealistic images from new perspectives.

1.9.5 Input Image Collection

The first step in the NeRF pipeline is collecting a set of 2D images that capture the scene from various viewpoints. These images serve as the training dataset for the neural network. Along with the images, camera parameters such as position, orientation, and intrinsic properties (e.g., focal length) are recorded to provide the network with the necessary context for reconstructing the scene in 3D.

For optimal performance, the training dataset should cover the scene from a wide range of angles to avoid missing important details. Sparse or incomplete coverage can lead to gaps in the reconstructed scene or visual artifacts in the synthesized views. In VR and AR applications, these input images may be captured using panoramic cameras, drones, or even depth sensors to improve the accuracy of the training data.

1.9.6 Neural Network Training

Once the dataset is prepared, the neural network is trained to approximate the radiance field of the scene (Figure 1.6). The training process involves optimizing the network's weights by minimizing the loss function, which measures the difference between the rendered outputs and the ground truth 2D images. The network employs a fully connected multilayer perceptron (MLP) architecture, where each layer applies non-linear transformations to learn complex relationships between spatial coordinates and color values.

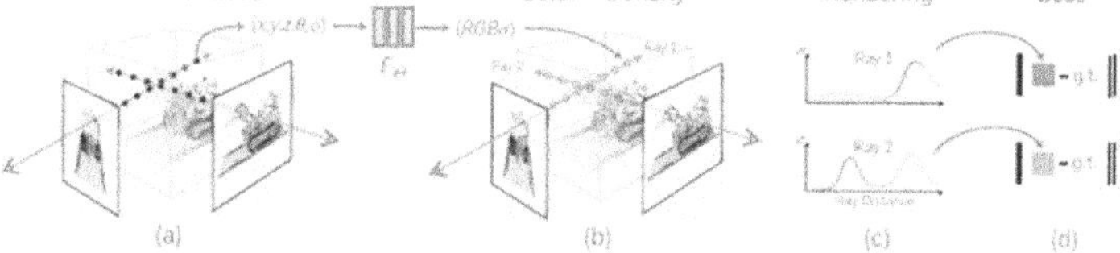

Figure 1.6 NeRF pipeline [11].

During training, the network predicts the color and density values for each 3D point sampled along a ray projected from the camera through the scene. By integrating these values along each ray, the system reconstructs the final pixel color for the corresponding image. This process is repeated for thousands of training iterations until the network can accurately reproduce the input images from the training dataset.

1.9.7 Rendering Novel Views

After training, the network can render novel views of the scene by sampling rays from new camera positions. For each ray, the network predicts the color and density values at multiple points along the ray's path. These values are integrated to compute the final pixel color for the synthesized image. Unlike traditional rendering, which relies on predefined geometric meshes and lighting equations, NeRF generates the scene directly from its learned representation.

The rendering process is computationally intensive, as it requires evaluating the neural network for thousands of points per ray and millions of rays per frame. However, recent optimizations, such as hierarchical volume sampling and factorized neural networks, have significantly improved NeRF's rendering speed, making it more feasible for real-time applications.

1.10 COMPARISON WITH TRADITIONAL RENDERING TECHNIQUES

NeRF's data-driven approach to 3D scene reconstruction and rendering offers several advantages over traditional polygon-based rendering methods, but it also introduces new challenges. This section compares the two approaches in terms of realism, computational efficiency, and flexibility.

1.10.1 Advantages of NeRF Over Polygon-Based Rendering

1. **Photorealistic Output:** NeRF can produce highly realistic images with complex lighting effects, such as global illumination, soft shadows, and specular reflections. Traditional rasterization-based rendering methods require additional techniques, such as shadow mapping and reflection probes, to approximate these effects, often resulting in visual artifacts. Ray tracing can achieve similar levels of realism but at a significantly higher computational cost.
2. **Implicit Scene Representation:** Unlike traditional methods that use explicit geometric meshes, NeRF encodes the entire scene as a neural network. This implicit representation allows NeRF to model complex, non-uniform structures such as hair, foliage, and semi-transparent materials more effectively than polygonal meshes.

3. **View-Consistent Rendering:** NeRF generates view-consistent images from any perspective, eliminating issues such as popping and discontinuities that can occur with LOD-based optimizations in traditional rendering. This is particularly important in VR/AR applications, where abrupt changes in detail can break immersion.
4. **Efficient Handling of Scene Complexity:** NeRF's learned representation scales more efficiently with scene complexity compared to polygonal rendering, which requires large amounts of memory to store high-resolution textures and detailed meshes.

1.10.2 Limitations of NeRF

1. **Computational Intensity:** Despite recent optimizations, NeRF's rendering process remains computationally expensive, requiring significant GPU resources to evaluate the neural network for millions of points per frame. This makes real-time applications challenging on standalone VR/AR headsets with limited processing power.
2. **Training Time:** Training a NeRF model can take hours or even days, depending on the size of the dataset and the complexity of the scene. This is a significant drawback compared to traditional rendering methods, which do not require pre-training.
3. **Static Scene Assumption:** Early implementations of NeRF are designed for static scenes and encounter challenges in handling dynamic elements, such as moving objects or changing light sources. While research into dynamic NeRF models is ongoing, traditional rendering techniques still have an advantage in interactive environments.
4. **Techniques in Neural Rendering for VR/AR:** Neural rendering has redefined how VR/AR systems generate high-fidelity, interactive visual content. Figure 1.7 shows an AR view of the NeRF model. While neural rendering offers photorealistic visuals and improved scene realism compared to traditional rendering, achieving real-time performance remains a significant challenge due to the computational demands of deep neural networks. However, innovative techniques, such as Gaussian splatting, dynamic scene generation, and hardware-driven optimizations, are paving the way for faster, more efficient neural rendering. This section explores these key techniques and their contributions to improving neural rendering in VR/AR applications.

1.10.3 Gaussian Splatting for Real-Time Performance

1.10.3.1 Overview of Gaussian Splatting

Gaussian splatting [17] is a novel rendering technique that accelerates neural rendering by approximating 3D scene elements as Gaussian [23] blobs rather than explicit geometric surfaces or pixel-based point clouds. In this

Figure 1.7 NeRF in HoloLens 2.

approach, a scene is represented as a set of 3D Gaussians, each defined by its center point, covariance matrix (representing shape and orientation), and color. These Gaussians are rendered directly as elliptical splats that blend seamlessly based on their overlapping regions, creating a continuous and smooth representation of surfaces without the need for complex mesh connectivity.

Gaussian splatting effectively replaces traditional mesh-based scene representations with a probabilistic, volumetric distribution of points, allowing neural rendering to operate more efficiently by minimizing the number of computations required for each frame. This approach bypasses the need to process millions of polygons and enables direct sampling of 3D points during rendering, significantly reducing memory overhead and improving performance.

1.10.3.2 Advantages for Faster Rendering

One of the primary advantages of Gaussian splatting is its ability to achieve real-time performance even in complex scenes. By representing surfaces as overlapping Gaussians, this technique simplifies the rendering pipeline and avoids the need for costly ray-tracing operations or complex neural network evaluations at every pixel.

- **Memory Efficiency:** Since Gaussians can represent smooth, continuous surfaces with fewer data points than traditional mesh-based methods, this technique significantly reduces memory usage. In scenes with detailed textures and fine-grained geometry, Gaussian splatting requires fewer parameters to maintain visual fidelity.

- **Dynamic Lighting Compatibility:** Gaussian splatting integrates well with dynamic lighting models, allowing for soft shadows and ambient occlusion effects to be computed more efficiently. Instead of recalculating ray-traced reflections or shadows for each frame, the Gaussians can store precomputed approximations of how light interacts with different parts of the scene.
- **Temporal Coherence:** In VR/AR applications, maintaining temporal coherence – that is, consistent visuals across time as the user moves – prevents distracting visual artifacts. Gaussian splatting's probabilistic blending of neighboring splats avoids "popping" effects that occur in LOD-based systems, ensuring a smoother transition between perspectives.

By reducing the complexity of the rendering process, Gaussian Splatting brings neural rendering closer to achieving the low latency and high frame rates required for interactive VR/AR experiences on standalone devices

1.11 DYNAMIC SCENE GENERATION AND MOTION HANDLING

1.11.1 Techniques for Rendering Dynamic, Interactive Environments in Real Time

While neural rendering methods such as NeRF excel at reconstructing static scenes, real-time VR/AR applications often involve dynamic environments where objects move, change shape, or interact with the user. Achieving real-time performance in such scenarios requires specialized techniques for motion handling, dynamic object synthesis, and scene updates.

1. **Dynamic NeRF (D-NeRF):** Dynamic NeRF [18] extends the capabilities of static NeRF models by incorporating time as an additional dimension in the neural network. Instead of mapping only 3D spatial coordinates to color and density values, D-NeRF maps (x, y, z, t) coordinates, where t represents time. This allows the network to capture and render temporal changes, such as moving objects or animated characters. However, the inclusion of the time dimension increases the complexity of the model, requiring further optimization to maintain real-time performance.
2. **Optical Flow-Based Motion Compensation:** Another approach to handling motion in neural rendering is the use of optical flow, which estimates the movement of pixels between consecutive frames. By predicting how points in the scene shift over time, neural rendering systems can preemptively adjust the scene's structure and appearance,

reducing the need to reprocess the entire scene. This technique is particularly useful for VR/AR applications, where user head movements and interactions occur at high speeds, requiring seamless visual updates to avoid latency-related artifacts.

3. **Implicit Warp Fields:** Implicit warp fields are neural networks that learn how objects deform or move across time. These networks predict how the underlying 3D scene structure changes in response to external forces or user interactions, enabling the system to render deformable objects, such as flexible characters or flowing water. By learning a mapping function from an object's initial state to its deformed state, warp fields can reduce the computational cost of simulating complex physical interactions in real-time VR/AR environments.

1.11.2 Real-Time Scene Composition for VR/AR

In interactive VR/AR applications, the user's experience is shaped by how quickly and accurately the scene adapts to their actions and surroundings. Techniques such as *hybrid neural rendering* combine neural networks with traditional rasterization or ray tracing to improve the efficiency of dynamic scene rendering. For example, static background elements can be rendered using rasterization, while dynamic foreground elements – such as avatars or interactive objects – are rendered using neural networks to ensure photorealistic animations and interactions.

1.12 HARDWARE-DRIVEN OPTIMIZATIONS

1.12.1 How Advanced Hardware Supports Neural Rendering

The effectiveness of neural rendering depends not only on the efficiency of the algorithms but also on the capabilities of the underlying hardware. Modern VR/AR headsets, such as the Apple Vision Pro and Meta Quest 3, are equipped with specialized hardware components designed to accelerate neural network computations and improve rendering performance.

1. **Neural Processing Units:** NPUs are specialized processors designed to handle the parallel computations involved in neural network inference. Devices such as Meta Quest 3 feature integrated NPUs that can perform matrix multiplications and activation functions more efficiently than traditional CPUs and GPUs. By offloading neural rendering tasks to NPUs, these devices achieve lower latency and higher frame rates, making real-time neural rendering feasible for standalone headsets.
2. **SoC Architectures:** The Apple Vision Pro's M2 chip and R1 coprocessor exemplify the advantages of SoC architectures, which integrate

CPUs, GPUs, and NPUs into a single chip. This integration reduces communication delays between processing units and improves energy efficiency, allowing the device to run complex neural rendering models without overheating or draining the battery quickly. The R1 coprocessor in the Vision Pro is specifically optimized for sensor fusion, combining data from multiple cameras and depth sensors to create accurate spatial maps for neural rendering in mixed reality (MR) environments.

3. **Foveated Rendering:** Foveated rendering is a hardware-driven optimization that uses eye-tracking sensors to determine the user's focal point and prioritize rendering detail in that area. The Apple Vision Pro, for example, renders the region in the center of the user's gaze at maximum resolution while reducing detail in peripheral regions. This technique significantly reduces the number of computations needed per frame, freeing up resources for more complex neural rendering tasks.
4. **Depth Sensing and SLAM:** Advanced depth sensors and SLAM algorithms enable VR/AR devices to build detailed 3D maps of the physical environment in real time. This spatial mapping data can be used by neural rendering systems to accurately position virtual objects within the user's surroundings and ensure that interactions between real and virtual elements appear seamless.

1.12.2 Camera Systems and Sensor Fusion

The Meta Quest 3 and Apple Vision Pro feature multiple cameras and sensors that capture RGB, infrared, and depth data to improve spatial awareness and scene reconstruction. This sensor fusion allows neural rendering systems to account for real-world changes, such as lighting variations or moving objects, ensuring that virtual content remains anchored to the physical environment.

1.13 NEURAL RENDERING FOR HUMAN RECONSTRUCTION AND INTERACTION

One of the most promising applications of neural rendering in VR and AR is in the creation of lifelike human avatars and real-time interactions. Human reconstruction and motion synthesis play a pivotal role in immersive experiences such as virtual meetings, gaming, and medical applications. By leveraging neural networks, these systems can create highly realistic representations of people, capturing subtle details such as facial expressions, body movements, and pose dynamics. This section explores techniques for real-time human avatar creation, 3D pose estimation[23]

[34], and a practical case study in telerehabilitation, demonstrating how neural rendering enhances human-centered VR/AR applications.

1.13.1 Real-Time 3D Human Avatar Creation

Real-time 3D human avatar creation involves generating realistic, animated digital representations of people within VR and AR environments. Traditional approaches use methods such as mesh-based modeling and motion capture, which, although accurate, require extensive preprocessing and expensive equipment. Neural rendering simplifies this process by learning from 2D image data and generating avatars directly through neural network inference.

1.14 TECHNIQUES FOR CREATING REALISTIC HUMAN AVATARS

1. **Neural Body Models (e.g., SMPL):** One popular approach involves the use of parametric models, such as the Skinned Multi-Person Linear (SMPL) [19] model. SMPL is a learned model of the human body that encodes shape and pose variations in a compact, differentiable representation. By training neural networks to control the parameters of SMPL, neural rendering systems can generate realistic human bodies with diverse appearances and dynamic motions.
2. **Implicit Surface Representations:** Implicit neural representations, such as signed distance functions (SDFs) or volumetric neural fields, are used to represent the 3D structure of the human body. These methods encode the surface geometry of the avatar as a continuous function, allowing for smooth and detailed reconstructions without explicit polygonal meshes. This approach is particularly effective for handling complex details such as clothing folds, hair, and hands.
3. **Real-Time Facial Animation:** Neural rendering excels at generating realistic facial animations by learning to map facial landmarks or expression coefficients to rendered outputs. Techniques such as neural face rendering (NFR) [20] combine 3D morphable models (3DMM) [21] with neural networks to synthesize photorealistic expressions and emotions in real time, even from low-resolution inputs such as webcam footage. This capability is essential for enhancing non-verbal communication in virtual interactions.

By using these techniques, neural rendering systems can create avatars that move and respond naturally to user inputs, making virtual representations of people more immersive and believable.

1.15 CHALLENGES AND FUTURE DIRECTIONS

While neural rendering has demonstrated significant advancements, several challenges must be addressed to fully realize its potential in real-time VR/AR applications. This section outlines key challenges and emerging trends in the field.

1.15.1 Hardware Constraints

1.15.1.1 Limitations of Standalone VR/AR Devices for Real-Time Neural Rendering

Standalone VR/AR devices, such as the Meta Quest 3 and Apple Vision Pro, offer portability and convenience but are limited by their computational power and battery life. Neural rendering is highly resource-intensive, requiring large-scale matrix operations and neural network inference for every frame.

1. **Memory Limitations:** Neural rendering models, such as NeRF, require significant memory to store volumetric data and intermediate computations. Standalone headsets often have limited GPU memory compared to desktop systems, which can result in reduced visual fidelity.
2. **Thermal Constraints:** Running neural network inference at high frame rates generates substantial heat, which can cause performance throttling and reduced battery life. Advanced cooling systems are needed to support sustained performance.

To address these constraints, hardware optimizations such as dedicated NPUs and SoC architectures have been integrated into modern headsets. However, further advancements are needed to balance performance with energy efficiency.

1.15.2 Latency and Synchronization Issues

1.15.2.1 Impact of Rendering Delays on User Experience

In VR/AR, even minor delays in rendering can cause noticeable desynchronization between user actions and visual updates, leading to motion sickness and discomfort.

1. **Frame Latency:** Neural rendering pipelines introduce latency due to the time required to evaluate neural networks for millions of points per frame. This can result in lag, particularly during fast head movements or interactions with dynamic objects.

2. **Synchronization across Devices:** In collaborative VR/AR applications, ensuring that multiple participants perceive the same environment in real time requires precise synchronization of avatar positions, audio, and scene updates. Any delay in transmitting neural-rendered data can break the sense of co-presence.

To mitigate these issues, low-latency neural architectures, such as efficient neural encoders and hierarchical sampling, are being developed to reduce inference times and maintain frame rates above 90 FPS.

1.16 CONCLUSION

Recap of Key Discussions and Takeaways
This chapter explored the applications of neural rendering in healthcare, collaborative telepresence, and entertainment, highlighting its ability to create photorealistic avatars, dynamic virtual scenes, and immersive gaming experiences. The discussion also addressed the challenges of real-time neural rendering, including hardware limitations and latency issues.

Emphasis on the Transformative Potential of Neural Rendering
Neural rendering has the potential to redefine VR/AR experiences by bridging the gap between pre-rendered visualizations and real-time interactivity. By leveraging neural networks, these systems can generate lifelike environments that respond dynamically to user actions, enhancing immersion and engagement across various applications.

Encouragement for Further Research and Development
To fully unlock the capabilities of neural rendering, continued research is needed in areas such as federated learning, hardware/software co-design, and dynamic scene synthesis. As advancements in neural network optimization and hardware acceleration continue, neural rendering will play a critical role in shaping the future of VR/AR, making it more accessible, efficient, and transformative.

REFERENCES

1. Fracaro, S. G., Tehreem, Y., Toyoda, R., Gallagher, T., Glassey, J., Bernaerts, K., & Wilk, M. (2024). Benefits and impact of emergency training in a VR environment. *Education for Chemical Engineers*.
2. Gutierrez, N. (2023). The ballad of morton heilig: on VR's mythic past. *JCMS: Journal of Cinema and Media Studies*, *62*(3), 86–106.
3. Jacobsen, S. D. (2019). Global democracy's sword of damocles–the politics of science and survival, or the dance of the polichinelles. *People*, *2020*, 2021.

4. Creagh, H. (2003, September). Cave automatic virtual environment. In *Proceedings: Electrical Insulation Conference and Electrical Manufacturing and Coil Winding Technology Conference (Cat. No. 03CH37480)* (pp. 499–504). IEEE.
5. Desai, P. R., Desai, P. N., Ajmera, K. D., & Mehta, K. (2014). A review paper on oculus rift-a virtual reality headset. *arXiv preprint arXiv:1408.1173.*
6. Borges, M., Symington, A., Coltin, B., Smith, T., & Ventura, R. (2018, October). HTC vive: Analysis and accuracy improvement. In *2018 IEEE/RSJ International Conference on Intelligent Robots and Systems (IROS)* (pp. 2610–2615). IEEE.
7. Park, S., Bokijonov, S., & Choi, Y. (2021). Review of Microsoft HoloLens applications over the past five years. *Applied Sciences, 11*(16), 7259.
8. Caruso, T. J., Hess, O., Roy, K., Wang, E., Rodriguez, S., Palivathukal, C., & Haber, N. (2021). Integrated eye tracking on magic leap one during augmented reality medical simulation: a technical report. *BMJ Simulation & Technology Enhanced Learning*, 7(5), 431.
9. Aros, M., Tyger, C. L., & Chaparro, B. S. (2024, June). Unraveling the Meta Quest 3: An Out-of-Box Experience of the Future of Mixed Reality Headsets. In *International Conference on Human-Computer Interaction* (pp. 3–8). Springer Nature Switzerland.
10. Masalkhi, M., Waisberg, E., Ong, J., Zaman, N., Sarker, P., Lee, A. G., & Tavakkoli, A. (2023). Apple vision pro for ophthalmology and medicine. *Annals of Biomedical Engineering, 51*(12), 2643–2646.
11. Mildenhall, B., Srinivasan, P. P., Tancik, M., Barron, J. T., Ramamoorthi, R., & Ng, R. (2021). Nerf: Representing scenes as neural radiance fields for view synthesis. *Communications of the ACM, 65*(1), 99–106.
12. Han, D., & Yoo, H. J. (2023). *On-Chip Training NPU-Algorithm, Architecture and SoC Design*. Springer.
13. Park, S., Bokijonov, S., & Choi, Y. (2021). Review of Microsoft HoloLens applications over the past five years. *Applied Sciences, 11*(16), 7259.
14. Bishop, G., & Weimer, D. M. (1986). Fast phong shading. *ACM Siggraph Computer Graphics, 20*(4), 103–106.
15. Foley, J. D., Van Dam, A., Feiner, S. K., Hughes, J. F., & Phillips, R. L. (1994). *Introduction to Computer Graphics* (Vol. 55). Addison-Wesley.
16. Seo, J., Kim, G. J., & Kang, K. C. (1999, December). Levels of detail (LOD) engineering of VR objects. In *Proceedings of the ACM Symposium on Virtual Reality Software and Technology* (pp. 104–110).
17. Tamke, M. (2005, April). Baking light: Global illumination in VR environments as architectural design tool. In *CAADRIA 2005 in Proceedings of the 10th International Conference on Computer Aided Architectural Design Research in Asia New Delhi (India)* (Vol. 2, pp. 214–228).
18. Pumarola, A., Corona, E., Pons-Moll, G., & Moreno-Noguer, F. (2021). D-nerf: Neural radiance fields for dynamic scenes. In *Proceedings of the IEEE/CVF Conference on Computer Vision and Pattern Recognition* (pp. 10318–10327).
19. Loper, M., Mahmood, N., Romero, J., Pons-Moll, G., & Black, M. J. (2023). SMPL: A skinned multi-person linear model. In *Seminal Graphics Papers: Pushing the Boundaries* (Vol. 2 pp. 851–866).

20. Qin, D., Saito, J., Aigerman, N., Groueix, T., & Komura, T. (2023, July). Neural face rigging for animating and retargeting facial meshes in the wild. In *ACM SIGGRAPH 2023 Conference Proceedings* (pp. 1–11).
21. Booth, J., Roussos, A., Ponniah, A., Dunaway, D., & Zafeiriou, S. (2018). Large scale 3D morphable models. *International Journal of Computer Vision*, *126*(2), 233–254.
22. Habgood, M. J., Wilson, D., Moore, D., & Alapont, S. (2017, October). HCI lessons from PlayStation VR. In *Extended Abstracts Publication of the Annual Symposium on Computer-Human Interaction in Play* (pp. 125–135).
23. Chen, G., & Wang, W. (2024). A survey on 3d Gaussian splatting. *arXiv preprint arXiv:2401.03890.*
24. Osokin, D. (2018). Real-time 2d multi-person pose estimation on CPU: Lightweight openpose. *arXiv preprint arXiv:1811.12004.*
25. Lugaresi, C., Tang, J., Nash, H., McClanahan, C., Uboweja, E., Hays, M., ... Grundmann, M. (2019). Mediapipe: A framework for building perception pipelines. *arXiv preprint arXiv:1906.08172.*

Chapter 2

3D Reconstruction of Metaverse Objects Using a Single Photograph

Lifan Xu and Shunqiao Sun

2.1 INTRODUCTION

The concept of social attributes stands as a pivotal characteristic in the Metaverse, encompassing immersive interactions with entities bridging the real and digital realms. The creation of virtual entities from images serves as a straightforward and efficient method to enhance user experiences. Despite the existing capability of certain devices, such as the latest iPhones and Android devices, to capture depth information in images for direct mapping of corresponding virtual entities, these devices are not universally accessible due to their high cost. The technique of reconstructing the 3D geometry and structure of an object from one or multiple images presents an intriguing prospect. The primary challenge in this context lies in the ambiguity arising from the absence of depth and semantic information. For human body reconstruction, a robust reconstruction technique should fulfill several criteria, including high resolution, efficiency, accuracy, and adaptability to various poses and clothing styles on different bodies. Similarly, for reconstructing 3D objects from a single 2D image, the reconstruction method needs to be robust and can be accurately reconstructed in the face of different input image angles or topologies.

2.2 HUMAN BODY RECONSTRUCTION

To precisely reconstruct 3D human bodies from a single image, the objectives are maintaining clothing textures and ensuring robustness to challenging poses. Based on their 3D representations, the methods can be broadly categorized into two groups: parametric, which includes HMD and Tex2Shape [1, 2], and non-parametric, encompassing voxel grids, silhouettes, implicit fields, and depth map methods. In contrast to parametric methods, non-parametric approaches exhibit the capability to construct 3D bodies with challenging poses and clothing textures but cannot construct objects precisely.

 DOI: 10.1201/9781032663609-3

The parametric model-conditioned implicit representation (PaMIR) [3] integrates the skinned multi-person linear model (SMPL) with an implicit surface function to achieve accurate reconstruction of human bodies with clothing textures. The system diagram is illustrated in Figure 2.1. SMPL represents a model for body shape and pose, incorporating shape variations dependent on identity. The model breaks down body shape into identity-dependent and pose-dependent components [4]. Traditional skinning equations, defined by pose and shape parameters, enable the fitting of 3D representations of the human body. When presented with an input image, the initial step involves estimating a corresponding SMPL model in the SMPL estimation phase. Subsequently, the image and the SMPL model are transformed into a feature map using a 2D encoder E_I and a feature volume using a 3D encoder E_V, respectively. Conditioned on the sampled feature vectors obtained during the feature sampling step, the final phase involves determining the implicit function value for each 3D point.

The method attains state-of-the-art performance in image-based 3D human reconstruction, particularly in scenarios involving challenging poses and various clothing types. The proposed methodology incorporates an implicit field for the construction of a 3D surface. A deep implicit function, implemented as a neural network, is employed to delineate whether a 3D point resides inside or outside a given 3D model. Introducing a conditioning variable serves to specify the model, expressed as

$$f_\theta(D,x):\mathbb{R}^3\times\chi\mapsto[0,1] \tag{2.1}$$

Here, f_θ represents the network mapping function, $D \in R^3$ denotes a point in 3D, and $x \in \chi$ is a condition variable defined by specific tasks. This approach eliminates the need for explicit storage of a surface, resulting in memory efficiency. To leverage the advantages of both parametric and non-parametric methods, a pixel-aligned implicit function is given as

$$f_\theta^p(d)=f_\theta\left(F_I(d),F_V(D)\right)^T \tag{2.2}$$

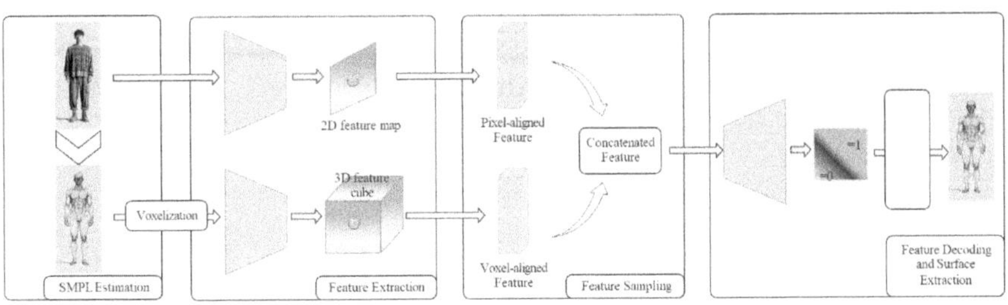

Figure 2.1 System diagram of the PaMIR.

For a 3D point D, $d = \pi\ (D)$ represents its 2D projection. $F_I(d) = E_I(I(d))$ is the image feature at D. F_V is the feature column at point D. The generation of the feature volume F_V begins with the conversion of the SMPL mesh into an occupancy volume (O_v) through the process of mesh voxelization. Subsequently, this volume is encoded using a 3D encoder E_v, described as $F_V = E_V\ (O_v)$. During the training process, the SMPL model estimation is based on a single image, and therefore for body poses, depth ambiguity is inevitable because the alignment between the estimation and the ground truth along depth axes cannot be guaranteed. To address this issue, a novel reconstruction loss is given by

$$\mathcal{L} = \frac{1}{N}\sum_{i=1}^{N}\left|f_\theta^p\left(d_i + \Delta_{d_i}\right) - f_\theta^{*p}\left(d_i\right)\right| \tag{2.3}$$

where we assume N points are sampled, and the translation compensation number $\Delta_{d_i} = (0, 0, \Delta_{z_i}\ )$ is given by

$$\Delta_{z_i} = \sum_{j\in\mathcal{N}(i)} \frac{\alpha_{j\to i}}{\alpha_i}\left(Z\left(v_j\right) - Z\left(v_j^*\right)\right) \tag{2.4}$$

In the given expression, $\mathcal{N}(i)$ represents the nearest set of SMPL vertices for d_i, and $|\ \mathcal{N}(i)| = 4$. $\alpha_{j\to i}$ signifies the corresponding blending weight, α_i denotes the weight normalizer, v_j represents the j-th vertex of the predicted MSPL model, and v_j^* is the corresponding ground-truth SMPL model vertex.

In Equation (2.3), the training process relies on SMPL estimation, and any inconsistency between the prediction model and the ground truth leads to reconstruction errors. To correct the error, an improved accuracy can be obtained by minimizing

$$\left(\beta^*, \sigma^*\right) = {}^{argmin}_{\beta,\sigma}\mathcal{L}_I + \lambda_R\mathcal{L}_R \tag{2.5}$$

where bady fitting loss $\mathcal{L}_I$ is to narrow down the difference between the predicted implicit function and the ground-truth SMPL model, and the regularization term $\mathcal{L}_R$ is to penalize the difference between the initial prediction model parameter (β_0, σ_0) generated by (Graph-CNN-based Mesh Regression GCMR) algorithm and the truth parameters (β^* , σ^*). With Q SMPL vertices, $\mathcal{L}_I$ is given by

$$\mathcal{L}_I = \frac{1}{Q}\sum_{j=1}^{Q} P\left(f_\theta^p\left(v_j\right) - 0.5\right) \tag{2.6}$$

Where the penalty loss $P\ (*)$ is defined as

$$P\left(v_j\right)\begin{cases} \left|v_j\right|, v_j \geq 0, \\ \dfrac{1}{\eta\left|v_j\right|}, v_j < 0 \end{cases} \tag{2.7}$$

The regularization term $\mathcal{L}_R$ is

$$\mathcal{L}_R = \left|\beta - \beta_0\right|_2^2 + \left|\sigma - \sigma_0\right|_2^2 \tag{2.8}$$

In the pursuit of achieving state-of-the-art reconstruction quality, the researchers have meticulously compiled a dataset comprising 1,000 high-quality textured human scans. This dataset captures a diverse spectrum of variations, encompassing distinct clothing styles, body shapes, and poses. Sourced attentively from Twindom, it serves as a comprehensive and varied resource for ongoing research endeavors.

The training process employs the Adam optimizer, incorporating a learning rate that undergoes decay across iterations. During the testing phase, the model requires only an RGB image as input and delivers both the parametric model and the reconstructed surface with texture. To assess generalization performance in single-image reconstruction, real-world full-body images were collected from both the DeepFashion dataset and the internet.

While the PaMIR method effectively combines parametric and non-parametric approaches for reconstructing 3D characters from 2D image input, it encounters significant challenges in reconstructing the backside of the character. This difficulty arises due to the complete loss of backside information in the 2D image.

The primary objective of [5] is to create a 3D human-shaped avatar using a single frontal image, as illustrated in the diagram shown in (Figure 2.2). This method not only accomplishes the reconstruction of the 3D human model but also provides a body texture map.

The initial step involves segmenting the human body shape using the cutting-edge human garment segmentation method. Following this, we apply a deformable parametric model called "SMPL" to conform to the segmented body shape. For the creation of the front geometry, we distort the depth map of the SMPL parametric model using correspondences established through silhouettes. Subsequently, we generate the back geometry using a back-culling rendering technique. In conclusion, we introduce a GAN network structure named InferGAN to recover the occluded back texture, training it on a combination of both real and synthetic images.

To implement a warping function for reconstructing the ultimate shape of the body, the initial step involves registering the SMPL silhouette to the human silhouette. The SMPL silhouette model, denoted as S after segmentation, comprises vertices $S_i \middle| i = 1, \cdots, v_S$, while the corresponding ground-truth human silhouette is represented as G, with vertices $G_i | i = 1, \cdots, v_G$. To ease the computation, an m-points uniform sampling is applied along S to obtain a sampling boundary set P and $P = \{p_1, p_2, \cdots, p_m\}$.The deform graph is utilized to represent the silhouette movement and hence solve a set of affine transformations $\mathbf{B} = \{B_j\}_{j=1}^{m}$ and $\mathbf{a} = \{a_j\}_{j=1}^{m}$. A minimization is used to solve the transformations, which is given by

$$\min \Upsilon_1 \Omega_1 + \Upsilon_2 \Omega_2 + \Omega_3 \tag{2.9}$$

where term Ω_1 preserves the rigidity property of the rotation matrix and is therefore defined as

$$\Omega_1 = \sum_{j=1}^{m} B_j^T B_j - \boldsymbol{I}_F^2 + \left(\det\left(B_j\right) - 1\right)^2 \tag{2.10}$$

The second term, denoted as Ω_2, ensures spatial smoothness of the geometric deformation and is defined as

$$\Omega_2 = \sum_{j=1}^{m} \sum_{n=1}^{\mathcal{N}} B_j \left(p_n - p_j\right) + p_j + a_j - \left(p_n + a_n\right) \tag{2.11}$$

where $\mathcal{N}$ denotes the $\mathcal{N}$ nearest neighbors of the j-th node. The last term Ω_3 is given by

$$\Omega_3 = \sum_{j=1}^{m} \lambda_1 \hat{S}_j - G_{S_j}{}^2 + \lambda_2 \eta_j^T \left(\hat{S}_j - G_{S_j}\right)^2 \tag{2.12}$$

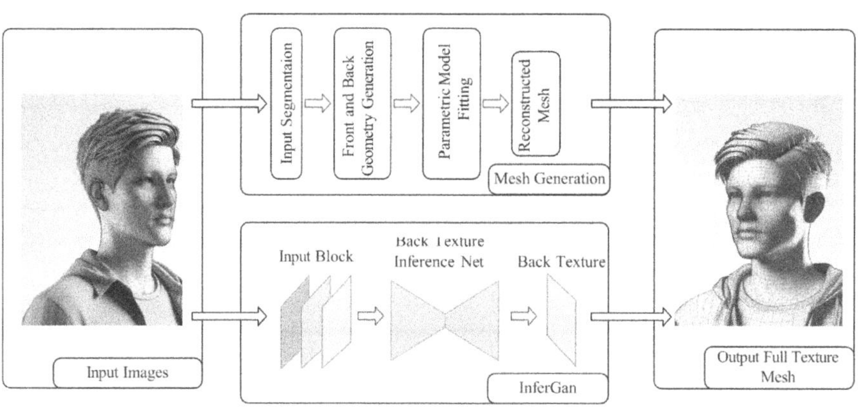

Figure 2.2 GAN-based texture inference system diagram.

where $\hat{S}_j = \sum_{j=1}^{m} \omega_j(S_j)\left[B_j(S_j - p_j) + b_j + a_j\right]$, G_{S_j} is the closest point of S_j on the ground truth silhouette, and η_j^T is the corresponding normal of G_{S_j}.

To construct the 3D object from a single image, the next thing is to assign a depth value to every point within the SMPL model. For each pixel inside the person mask, a depth value can be assigned based on the depth value of the corresponding point C(q) from the SMPL depth map:

$$Z(q) = ZSMPL(C(q)) \tag{2.13}$$

All the mentioned processes contribute to providing the front view of the human body, while the back view remains unknown. For a precise estimation, InferGAN is employed to transfer the input texture from one RGB image to another RGB image. This transfer is based on the input body semantic parsing map and the parsing map from the target view.

The proposal suggests employing 2D non-rigid registration to deform geometry from an initial estimate to the final reconstruction. In terms of texture extraction, it introduces the InferGAN, a mechanism capable of utilizing a frontal texture to infer textures at various view angles. This method not only achieves the reconstruction of complete geometry and intricate textures for a human body using just a single image but also acquires an animated model through the transfer of weights and joint locations from the parametric human model. The approach exhibits several limitations. First, it necessitates the human subject to face the camera to capture a frontal view effectively. Second, while the method can address partial occlusion cases, it requires a manual process for generating occluded regions, and it assumes that the subject's limbs and body do not intersect. Lastly, as SMPL is a skinned human body model, certain garment details are lost during the reconstruction process.

The idea of leveraging transformers for 3D reconstruction from a single 2D image is appealing for its potential to create lightweight and simplified models. This approach involves directly utilizing transformers to carry out the 3D reconstruction process, streamlining the overall model architecture.

K. Lin, L. Wang, and Z. Liu [6] introduce a novel technique called MEsh TRansfOrmer (METRO) (illustrated in Figure 2.3) for reconstructing 3D human pose and mesh vertices from a single image. METRO employs a transformer encoder to model interactions between vertices and joints collectively, generating simultaneous outputs for 3D joint coordinates and mesh vertices. In contrast to existing methods that rely on parametric mesh models such as SMPL, METRO does not depend on such models, making it easily extendable to other objects such as hands. The approach also relaxes mesh topology, allowing the transformer's self-attention mechanism to freely attend between any two vertices and enabling the learning of non-local relationships among mesh vertices and joints. The proposed masked vertex modeling enhances robustness and effectiveness in handling challenging scenarios, including partial occlusions.

The suggested framework comprises two modules: a convolutional neural network (CNN) and a multilayer transformer encoder. Initially, a CNN is employed to extract an image feature vector from the input image. The CNN undergoes pre-training on the ImageNet classification task. Specifically, a feature vector *V* is extracted from the last hidden layer, typically with a dimension of 2048. This extracted feature vector *V* is then fed into the transformer for the regression task. This generic design facilitates end-to-end training for human pose and mesh reconstruction. Additionally, the transformer can seamlessly leverage the advantages of large-scale pre-trained CNNs.

Subsequently, the multilayer transformer encoder receives the feature vector as input and concurrently produces the 3D coordinates of both body joints and mesh vertices. The model employs linear projections to diminish the dimensionality of the hidden embedding after each encoder layer. By incorporating multiple encoder layers, the model is perceived as conducting self-attentions and dimensionality reduction alternately. The ultimate output vectors from the transformer encoder represent the 3D coordinates of both joints and mesh vertices. In order to fully activate bidirectional attentions within its transformer encoder, a masked vertex modeling (MVM) is implemented for the regression task. This involves randomly masking a certain percentage of input queries, compelling the transformer to regress all joints and vertices despite the masking. To predict an output corresponding to a missing query, the model resorts to other relevant queries, akin to simulating occlusions where partial body parts are invisible. Consequently, MVM compels the transformer to regress 3D coordinates by considering

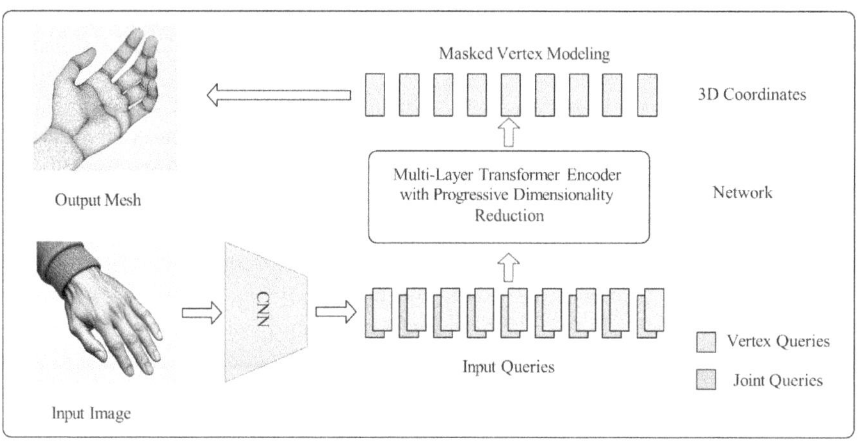

Figure 2.3 The MEsh TRansfOrmer (METRO) system diagram.

other pertinent vertices and joints, irrespective of their distances and mesh topology. This approach facilitates both short- and long-range interactions among joints and vertices, contributing to improved human body modeling.

For network training, multiple 2D and 3D datasets are utilized with a mixed-training strategy, including Human3.6M, 3DPW, and UP-3D, among others [7–10]. The training loss is composed as follows:

$$\mathcal{L} = \alpha_1\left(\mathcal{L}_B + \mathcal{L}_X + \hat{\mathcal{L}}_X\right) + \alpha_2 \acute{\mathcal{L}}_X \tag{2.14}$$

Here, α_1 and α_2 serve as indicators for 2D and 3D ground truths, $\mathcal{L}_B$ represents the L_1 loss between the predicted body joints $\hat{B}_j$ and the corresponding ground truth B_j, the L_1 loss of predicted body vertices $\hat{X}$ and the corresponding ground truth X, $\hat{\mathcal{L}}_X$ denotes the regressed 3D joints, and the 2D body joints projection loss is denoted by $\acute{\mathcal{L}}_X$.

Upon closer examination, the enhancements in performance primarily stem from the input-dependent non-local interactions learned in METRO. This capability allows predictions to be made based on crucial joints and vertices, irrespective of the mesh topology. Table 2.1 provides a comparison of different algorithms for human body reconstruction from a single image.

2.3 OBJECT RECONSTRUCTION

Certainly, reconstructing three-dimensional objects from a single image is a significant and important topic, particularly in the context of the virtual world. While character reconstruction is essential, extending the focus to include objects enhances the overall immersive experience. Techniques that enable the accurate and detailed reconstruction of both characters and objects from single images contribute to the realism and richness of virtual environments. This area of research plays a crucial role in applications such as virtual reality, gaming, and computer-generated imagery.

Table 2.1 A comparison of various algorithms for 3D human body reconstruction from a single image

Algorithm	*Model*	*Complexity*	*With Texture*	*Generalization*
PaMIR	Parametric &	High	Yes	Medium
InferGAN	Implicit function	High	Yes	Medium
METRO	Parametric &	Medium	No	High
	Implicit function			
	Implicit function			

Q. Huang, H. Wang, and V. Koltun introduce a novel method for the automatic 3D reconstruction of objects in internet images [11]. This approach focuses on reconstructing objects from a single view. The central concept involves jointly analyzing sets of images $I_N = \{I_1, I_2, \cdots, I_N\}$ depicting various objects, along with smaller collections of preexisting 3D models $S_M = \{S_1, S_2, \cdots, S_M\}$, to analyze and reconstruct the image. The collaborative analysis normalizes the formulaic optimization problem, ensuring stability in corresponding estimations, and effectively reproduces the appearance of the target without relying on traditional multi-view cues. For the joint analysis of image and shape, several assumptions are necessary, including the following: (1) The input shapes are consistently maintained at a uniform scale and orientation, with the underlying reflectional symmetry plane, if present, aligned along the x-plane. (2) Each shape is composed of one or more disconnected segments.

This technique initiates the process with the camera pose estimation for each input image in I_N. To estimate the camera pose, several sets are required. The first one is the label space set $\mathcal{B}=\{\mathbf{B}_1,\mathbf{B}_2,\cdots,\mathbf{B}_N\}$, comprising 360 camera poses associated with the images set I_N achieved by sampling the upper half of the viewing sphere using farthest point sampling [12]. The camera is directed toward the center of each sample. For every shape model within the set S_M, all poses from B have been utilized for image rendering, resulting in the composition of the rendering image set R. Thus, the camera pose estimation can be obtained by

$$\min_{\mathcal{B}} \sum_{i=1}^{N} \alpha_i(\boldsymbol{B}_i) + \sum_{I_i,I_j \in \mathcal{N}} \eta\left(\angle\left(\boldsymbol{B}_i,\boldsymbol{B}_j\right)\right)\gamma\left(I_i,I_j\right) \tag{2.15}$$

The unary potentials, denoted as $\alpha_i(\boldsymbol{B}_i)$, are determined by the distribution of camera poses from images in R that share similarities with I_i. The term N establishes connections between each image and its closest neighbors in the HOG descriptor space, following the methodology introduced by Dalal and Triggs [13]. The penalty term η corresponds to a truncated L_1 penalty, and $\angle(\boldsymbol{B}_i,\boldsymbol{B}_j)$ signifies the angle between the optical axes of $\boldsymbol{B}_i$ and $\boldsymbol{B}_j$. The weight $\gamma(I_i,I_j)$ fine-tunes the strength of the pairwise term based on the similarity of images I_i and I_j, defined in relation to their distance in the HOG space [13].

Subsequently, the task involves computing dense correspondences that establish links between images in set I_N and between images in I_N and those in R. This process entails iterating over patches in all images within I and identifying patches in other images that exhibit similarity based on appearance features. A weighted graph is then constructed, connecting these similar patches. The construction of the graph is followed by the application of a robust clustering algorithm, facilitating the extraction of patch clusters with dependable correspondences within each cluster. Leveraging these correspondences, we align all patches within each cluster by optimizing a joint non-rigid alignment objective. The outcome is highly accurate pixel-level correspondences within the identified patch clusters.

The obtained correspondences play a crucial role in linking each image segment with a corresponding shape part and transferring adjacency and symmetry information from shapes to images. An initial model is constructed for each image $I_i \in I_N$, incorporating shape parts associated with each segment in I_N . This initial model undergoes optimization to achieve projective alignment with image contours. The optimization objective relies on projective correspondences, the precision of which hinges on the accuracy of the camera parameters. Acknowledging the approximate nature of the initial camera pose estimation, intrinsic and extrinsic camera parameters are integrated into the optimization process. This refinement occurs concurrently with optimizing rigid transforms and non-rigid deformations for all segments. The object's configuration is regularized by incorporating symmetry and adjacency relationships inferred from the input shapes and images.

This research introduces a single-view reconstruction approach aimed at producing 3D models of objects found in web images. Despite the promising aspects of the approach, it is crucial to acknowledge certain limitations. These include a dependency on the availability of initial segmentations for input shapes, potential inaccuracies in image reconstruction when relevant parts are absent in the shape collection or incorrectly identified by computed correspondences, and a lack of scalability.

In response to the performance degradation caused by inputting the same images with different orders and the long-term memory loss of recurrent neural networks (RNNs), the multi-scale context-aware 3D object reconstruction has been introduced [14], as depicted in in Figure 2.4. This approach aims to address these challenges and enhance the efficiency of 3D object reconstruction.

Multiple images of the same object, albeit with different orders, are concurrently input into the network to generate individual coarse volumes. The network comprises multiple encoder–decoder pairs, sharing parameters to alleviate training efforts. Following this, a multi-scale context-aware fusion module is introduced to selectively choose high-quality reconstructions for distinct parts from all coarse 3D volumes. This process culminates in the creation of a fused 3D volume. A refiner is integrated to produce the final output to rectify inaccuracies in the recovered parts within the fused 3D volume.

The decision network is referred to as the multi-scale context-aware fusion part. The rationale behind this lies in the observation that an object

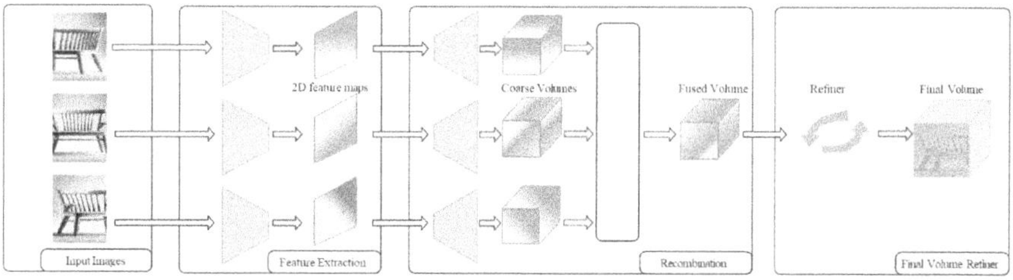

Figure 2.4 Enhanced Pix2Vox++ overview: This network reconstructs an object's 3D shape from multiple uncalibrated images, improving accuracy with more images. The encoder and decoder share weights across all views for efficient processing.

can be viewed from various perspectives. Consequently, the construction qualities of fusion-visible parts of an object from different angles are higher compared to a single view. For i-th coarse volume v_i^c, given a position (x_i, y_i, z_i) in the coarse volume, the normalized score $\hat{s}_i^{(x_i,y_i,z_i)}$ is given by

$$\hat{s}_i^{(x_i,y_i,z_i)} = \frac{exp(s_i^{(x_i,y_i,z_i)})}{\sum_{i=1}^{n} exp\left(s_i^{(x_i,y_i,z_i)}\right)} \tag{2.16}$$

where n represents the number of views. The fused voxel v_f is

$$v_f = \sum_{i=1}^{n} \hat{s}_i v_i^c \tag{2.17}$$

To train the network, the loss function can be defined as

$$\mathcal{L} = \frac{1}{N}\sum_{j=1}^{N}\left(p_j^* \log\left(p_j\right) + \left(1 - p_j^*\right) log\left(1 - p_j\right)\right) \tag{2.18}$$

where N is the number of voxels in the ground truth, p_j^* and p_j denote the ground truth occupancy and the predicted occupancy, respectively.

Aligned with the inherent characteristics of the network, it mandates multi-view inputs of the same object. To meet this requirement, a spectrum of datasets, including ShapeNet, Pix3D, and Things3D, has been integrated into the training regimen [15–17]. The intricacies of the training process unravel in a phased approach, commencing with the training of networks using single-view images as inputs for 250 epochs, with the exclusion of the multi-scale context-aware fusion component. Following this, the subsequent 100 epochs introduce multi-view images as inputs, contributing to the ongoing refinement of the networks. While this technique excels in reconstructing highly accurate objects, it is worth noting that the network components are intricate, and the datasets must encompass multi-view perspectives of the same object.

The proposed framework distinguishes itself from existing methods that directly fuse features from multi-view images. Instead, it opts to fuse the 3D volumes reconstructed from input images, a strategy that enhances the preservation of multi-view spatial constraints.

Nevertheless, image collections frequently encompass significant intraclass topological variations, including diverse perspectives of the same object or distinct entities within the same class (e.g., various types of tables and chairs). This presents a considerable challenge for reconstruction. Consequently, earlier approaches were either confined to estimating shapes and correspondences within classes lacking topological changes or concentrated solely on independently learning the shapes of each instance without establishing any correspondence. To tackle this issue, a topological-aware method has been introduced [18], and the corresponding diagram is provided in Figure 2.5. This method is designed to address challenges related to topology and contribute to improved performance in 3D reconstruction.

This technology comprises two primary components. In the initial step, a 3D point in the object space is mapped to its corresponding 3D point in the canonical space at the category level, accomplished through the DeformNet module. The canonical shape space serves as a unified representation, encompassing a diverse range of instances within the specified object category. The objective of DeformNet($\mathcal{F}$) is to acquire knowledge about dense 3D point deformations from the object space to the canonical space. To articulate it more precisely, when provided with an image I and a 3D point (x_i) in the object space, the deformation estimation task is formally defined as follows:

$$x_i + \mathcal{F}\left(x_i, I\right) = x_{c:3D} \tag{2.19}$$

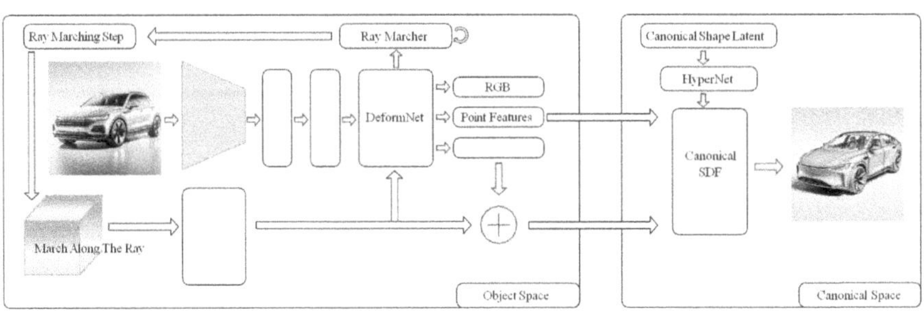

Figure 2.5 Topologically aware reconstruction from single-view system diagram.

In the expression, $x_{c:3D}$ denotes the corresponding point within the canonical space. The DeformNet ($\mathcal{F}$) function is akin to training a multilayer perceptron (MLP). To delve into the intricacies of the deformation field, fine-tuning the influence of the DeformNet module on the input image is crucial, which is accomplished through the hypernetwork. Initially, the input image is generated by the ImageNet pre-trained ResNet encoder to derive the foundational code. This computed code is then employed by the hypernetwork to predict the weights of the DeformNet MLP. The choice of a hypernetwork is deliberate, as it results in a more harmonious shape compared to the direct learning of MLP weights. Yet mastering deformation fields through MLP poses a formidable challenge due to the inherent continuity of MLPs. Although the continuous nature of MLPs aids in implicitly grasping 3D shapes, this very property hinders effective learning across object deformations. The issue lies in the fact that the deformation field between distinct topological objects may exhibit discontinuities. To surmount this hurdle and proficiently learn both deformation and shape fields, a level set method (LSM) comes into play [19]. LSM enables the topological merging and destruction of structures by representing any surface as a zero-horizontal crossover of a high-dimensional function. Consequently, MPL becomes adept at learning across diverse goals.

Following the initial step, where 3D points in the object space undergo deformation to their corresponding points in the canonical space, the subsequent phase involves learning the 3D shape in the form of a signed distance function (SDF) field [20]. To estimate the SDF values v_s for x_i, the associated higher-dimensional output $x_{c:HD}$ – formed by concatenating $x_{c:3D}$ with learned object space point features – is input into the canonical shape generator module F_c. Consequently, the value v_s is determined by the equation:

$$F_c(x_{c:HD}, C) = v_{s.} \tag{2.20}$$

Here, C denotes a canonical shape latent code within the hypernetwork. This code is instrumental in generating the weights for the shape generator and is jointly learned during the training process. The scene representation networks function as an LSTM-based differentiable renderer. This renderer operates through a ray marching procedure, wherein each marching step involves learning a depth estimate from the current 3D point along the current camera ray direction.

The datasets utilized for both training and evaluation encompass ShapeNet [21], Pix3D chairs [16], Pascal3D+ [22], and CUB-200-2011 [23]. Each training example comprises a cropped RGB image centered around the object, along with the corresponding segmentation map and camera pose. During inference, only the object image is required as input. The presented

Table 2.2 A comparison of various algorithms for 3D object reconstruction from a single image

Algorithm	Scalability	Complexity	Accuracy	Generalization
Joint Analysis	No	High	High	Medium
Pix2Vox++	No	Medium	High	Medium
TARS	No	Medium	Medium	Medium

method addresses the single-view reconstruction challenge through implicit learning, transforming diverse object instances into the learned class-specific average shape. The conversion of the three-dimensional deformation field to a high-dimensional deformation field demonstrates the topological awareness of the learned deformation field. Table 2.2 provides a comparison of different algorithms for object reconstruction from a single image.

2.4 CONCLUSION

Although the research on 3D models reconstructed from single images has been going on for quite some time, and many researchers have released refreshing new techniques, the field is still very challenging, and there are many directions worth exploring. The field of 3D reconstruction from a single image is dynamic, and researchers continue to explore various directions to advance the state-of-the-art. Some current and potential future research directions include the following:

A. *Improved Model Architectures:* Researchers are likely to focus on refining and innovating model architectures, combining traditional computer vision techniques with deep learning, and exploring novel network structures to enhance the accuracy and efficiency of 3D reconstruction.
B. *Handling Challenging Scenarios:* Addressing challenges such as occlusions, partial visibility, and handling diverse object categories is an ongoing area of research. Future work may involve developing models that are more robust in challenging real-world scenarios.
C. *Learning from Limited Data:* Extending 3D reconstruction models to generalize well with limited training data is crucial. Techniques for few-shot learning, domain adaptation, and transfer learning might be explored to make the models more applicable to diverse datasets.
D. *Real-Time and Scalability*: The development of real-time 3D reconstruction methods for applications such as augmented reality, robotics,

and virtual reality is a significant area of interest. Researchers may explore techniques to make the process more scalable and efficient.

E. *Combining Modalities:* Integrating information from multiple modalities, such as incorporating depth information from sensors or utilizing additional semantic cues, can potentially enhance the accuracy and reliability of 3D reconstruction models.

F. *Generalization across Domains:* Enhancing the ability of 3D reconstruction models to generalize across different domains, styles, and imaging conditions is crucial. This involves making models more adaptive and robust to variations in data.

G. *Ethical and Responsible AI* : As with any AI application, there is a growing emphasis on ensuring the ethical and responsible use of 3D reconstruction technology. Future research may explore ways to mitigate biases, ensure privacy, and address potential societal impacts.

REFERENCES

1. H. Zhu, X. Zuo, S. Wang, X. Cao, and R. Yang, "Detailed human shape estimation from a single image by hierarchical mesh deformation," Proceedings of the IEEE/CVF Conference on Computer Vision and Pattern Recognition, 2019, pp. 4491–4500.
2. T. Alldieck, G. Pons-Moll, C. Theobalt, and M. Magnor, "Tex2shape: Detailed full human body geometry from a single image," Proceedings of the IEEE/CVF International Conference on Computer Vision, 2019, pp. 2293–2303.
3. Z. Zheng, T. Yu, Y. Liu, and Q. Dai, "Pamir: Parametric model-conditioned implicit representation for image-based human reconstruction," *IEEE Transactions on Pattern Analysis and Machine Intelligence*, vol. 44, no. 6, pp. 3170–3184, 2021.
4. M. Loper, N. Mahmood, J. Romero, G. Pons-Moll, and M. J. Black, "Smpl: A skinned multi-person linear model," *Seminal Graphics Papers: Pushing the Boundaries*, vol. 2, 2023, pp. 851–866.
5. Z. Li, L. Chen, C. Liu, F. Zhang, Z. Li, Y. Gao, Y. Ha, C. Xu, S. Quan, and Y. Xu, "Animated 3d human avatars from a single image with gan-based texture inference," *Computers & Graphics*, vol. 95, pp. 81–91, 2021.
6. K. Lin, L. Wang, and Z. Liu, "End-to-end human pose and mesh reconstruction with transformers," Proceedings of the IEEE/CVF Conference on Computer Vision and Pattern Recognition, 2021, pp. 1954–1963.
7. C. Ionescu, D. Papava, V. Olaru, and C. Sminchisescu, "Human3. 6m: Large scale datasets and predictive methods for 3d human sensing in natural environments," *IEEE Transactions on Pattern Analysis and Machine Intelligence*, vol. 36, no. 7, pp. 1325–1339, 2013.

8. M. Andriluka, L. Pishchulin, P. Gehler, and B. Schiele, "2d human pose estimation: New benchmark and state of the art analysis," Proceedings of the IEEE Conference on Computer Vision and Pattern Recognition, 2014, pp. 3686–3693.
9. C. Lassner, J. Romero, M. Kiefel, F. Bogo, M. J. Black, and P. V. Gehler, "Unite the people: Closing the loop between 3d and 2d human representations," Proceedings of the IEEE Conference on Computer Vision and Pattern Recognition, 2017, pp. 6050–6059.
10. C. Zimmermann, D. Ceylan, J. Yang, B. Russell, M. Argus, and T. Brox, "Freihand: A dataset for markerless capture of hand pose and shape from single rgb images," Proceedings of the IEEE/CVF International Conference on Computer Vision, 2019, pp. 813–822.
11. Q. Huang, H. Wang, and V. Koltun, "Single-view reconstruction via joint analysis of image and shape collections." *ACM Transactions on Graphics*, vol. 34, no. 4, pp. 87–1, 2015.
12. T. F. Gonzalez, "Clustering to minimize the maximum intercluster distance," *Theoretical Computer Science*, vol. 38, pp. 293–306, 1985.
13. N. Dalal and B. Triggs, "Histograms of oriented gradients for human detection," 2005 IEEE Computer Society Conference on Computer Vision and Pattern Recognition (CVPR'05), vol. 1. IEEE, 2005, pp. 886–893.
14. H. Xie, H. Yao, S. Zhang, S. Zhou, and W. Sun, "Pix2vox++: Multi-scale context-aware 3d object reconstruction from single and multiple images," *International Journal of Computer Vision*, vol. 128, no. 12, pp. 2919–2935, 2020.
15. Z. Wu, S. Song, A. Khosla, F. Yu, L. Zhang, X. Tang, and J. Xiao, "3d shapenets: A deep representation for volumetric shapes," Proceedings of the IEEE Conference on Computer Vision and Pattern Recognition, 2015, pp. 1912–1920.
16. X. Sun, J. Wu, X. Zhang, Z. Zhang, C. Zhang, T. Xue, J. B. Tenenbaum, and W. T. Freeman, "Pix3d: Dataset and methods for single-image 3d shape modeling," Proceedings of the IEEE Conference on Computer Vision and Pattern Recognition, 2018, pp. 2974–2983.
17. S. Song, F. Yu, A. Zeng, A. X. Chang, M. Savva, and T. Funkhouser, "Semantic scene completion from a single depth image," Proceedings of the IEEE Conference on Computer Vision and Pattern Recognition, 2017, pp. 1746–1754.
18. S. Duggal and D. Pathak, "Topologically-aware deformation fields for single-view 3d reconstruction," Proceedings of the IEEE/CVF Conference on Computer Vision and Pattern Recognition, 2022, pp. 1536–1546.
19. S. Osher and R. P. Fedkiw, "Level set methods: An overview and some recent results," *Journal of Computational Physics*, vol. 169, no. 2, pp. 463–502, 2001.
20. C.-H. Lin, C. Wang, and S. Lucey, "Sdf-srn: Learning signed distance 3d object reconstruction from static images," *Advances in Neural Information Processing Systems*, vol. 33, pp. 11 453–11 464, 2020.

21. A. X. Chang, T. Funkhouser, L. Guibas, P. Hanrahan, Q. Huang, Z. Li, S. Savarese, M. Savva, S. Song, H. Su *et al.*, "Shapenet: An information-rich 3d model repository," *arXiv preprint arXiv:1512.03012*, 2015.
22. Y. Xiang, R. Mottaghi, and S. Savarese, "Beyond pascal: A benchmark for 3d object detection in the wild," IEEE Winter Conference on Applications of Computer Vision. IEEE, 2014, pp. 75–82.
23. P. Welinder, S. Branson, T. Mita, C. Wah, F. Schroff, S. Belongie, and P. Perona, *Caltech-ucsd Birds 200*, Technical Report CNS-TR-2010-001. California Institute of Tech-nology, 2010.

Chapter 3

Efficient and Scalable 3D Point Cloud Segmentation for the Metaverse

Yongzhi Yang and Kenneth Ricks

3.1 INTRODUCTION

The Metaverse is envisioned as a digital space that seamlessly merges the physical and virtual worlds [2]. For such an environment to function effectively, it needs a thorough understanding of 3D spatial data, enabling virtual elements to integrate naturally with real-world surroundings [4]. Point cloud segmentation is a key process in this context, as it identifies and classifies objects in a 3D scene, providing a detailed understanding of its structure [1]. Despite recent advancements, achieving accurate and efficient segmentation in various Metaverse scenarios remains a challenge [6, 7].

Traditional approaches, such as those based on voxel grids or projections, often lose important geometric details and are computationally expensive [1, 4]. Transformer-based architectures, which were originally designed for natural language processing, have shown promise in capturing both local and global contexts in 3D point clouds [2, 3]. Models such as KPConv, SalsaNext, Cylinder3D, and SphereFormer introduce new techniques to handle the limitations of earlier methods [1, 2, 4–6].

However, these models were designed primarily for structured environments, such as autonomous driving scenes, and may not perform well in the dynamic and diverse environments typical of the Metaverse [1, 4]. The interactive nature of the Metaverse also requires real-time processing capabilities, which many existing models struggle to provide [3, 6].

In this chapter, we present a framework that combines the strengths of several state-of-the-art transformer-based models for 3D point cloud segmentation [2–4]. Our contributions are as follows:

- We introduce a memory-efficient and scalable segmentation model that uses reversible transformer blocks, linear self-attention, and innovative architectural designs to maintain high performance without compromising on speed or efficiency [3].
- We test the proposed framework on both structured and unstructured datasets, demonstrating its ability to adapt and perform well across different types of environments [7–9].

DOI: 10.1201/9781032663609-4

- We discuss how our model can support real-time interactions and accurate scene understanding in the Metaverse, making it suitable for a variety of applications [2, 3].

The remainder of this chapter is structured as follows: The second section, "Related Work," reviews related work on point cloud segmentation and transformer-based models [1, 2, 4, 6]. The third, "Methodology," explains the methodology and implementation of the proposed framework. The fourth, "Experimental Results," provides experimental results and analysis. The fifth, "Application in the Metaverse," explores the potential impact of our model on Metaverse applications, and the concluding section "Conclusion and Future Work," concludes the chapter with future research directions [3].

3.2 RELATED WORK

With the rapid evolution of 3D point cloud segmentation techniques, it is crucial to understand the varying approaches and their impact on segmentation performance in diverse environments [1, 4]. Traditional methods, such as voxel-based and point-based models, have laid the foundation for segmentation tasks but often face limitations in handling complex spatial relationships and capturing long-range dependencies [1, 6]. Recently, transformer-based models have emerged as powerful alternatives, offering improved performance by using self-attention mechanisms [2, 3]. This section categorizes existing 3D point cloud segmentation methods into non-transformer models and transformer models, highlighting their key contributions, strengths, and limitations [1–4].

3.2.1 Non-Transformer Models for Point Cloud Segmentation

Non-transformer models for point cloud segmentation include projection-based, voxel-based, and point-based methods that rely on traditional deep learning architectures such as multilayer perceptrons (MLPs) and convolutional neural networks (CNNs) [1]. These methods are effective for capturing local features and handling geometric data but often struggle with modeling complex spatial relationships and capturing long-range dependencies due to the absence of self-attention mechanisms [1, 4, 6].

KPConv: KPConv introduces a unique convolutional operator specifically designed for point clouds. It uses rigid and deformable kernel points to adapt to local geometric structures and perform convolution directly on the point cloud [1]. Figure 3.1 shows the structure.

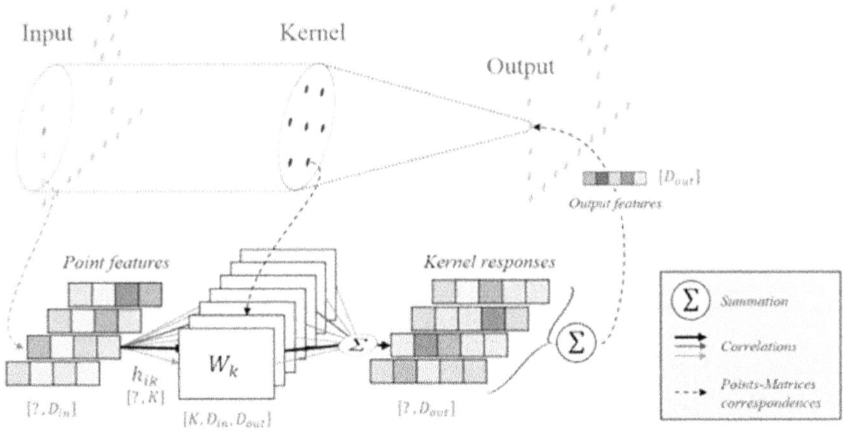

Figure 3.1 KPConv structure [1].

Cylinder3D: Cylinder3D proposes a cylindrical partitioning strategy to better handle the varying density of LiDAR point clouds, improving segmentation performance in driving scenarios [4]. Figure 3.2 shows the structure.

SalsaNext: SalsaNext is an extension of SalsaNet, designed to be lightweight and achieve real-time performance for LiDAR-based segmentation [6]. Figure 3.3 shows the structure.

3.2.2 Transformer Models for Point Cloud Segmentation

Transformer-based models have shown great promise in 3D vision tasks due to their ability to capture both local and global features using self-attention mechanisms [2, 3]. These models overcome the limitations of traditional architectures by modeling complex dependencies between points over varying distances [2, 3].

SphereFormer: SphereFormer uses a radial window self-attention mechanism to handle the sparsity and varying density of LiDAR data, making it suitable for both structured and unstructured environments [2]. Figure 3.4 shows the structure.

PReFormer: PReFormer addresses the high memory consumption of traditional transformers by introducing reversible transformer blocks and a linearized self-attention mechanism, allowing it to process larger inputs without excessive memory usage [3]. Figure 3.5 shows the structure.

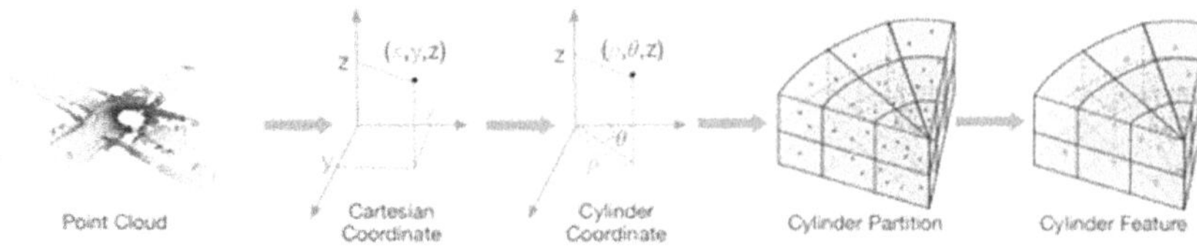

Figure 3.2 Cylinder3D structure [4].

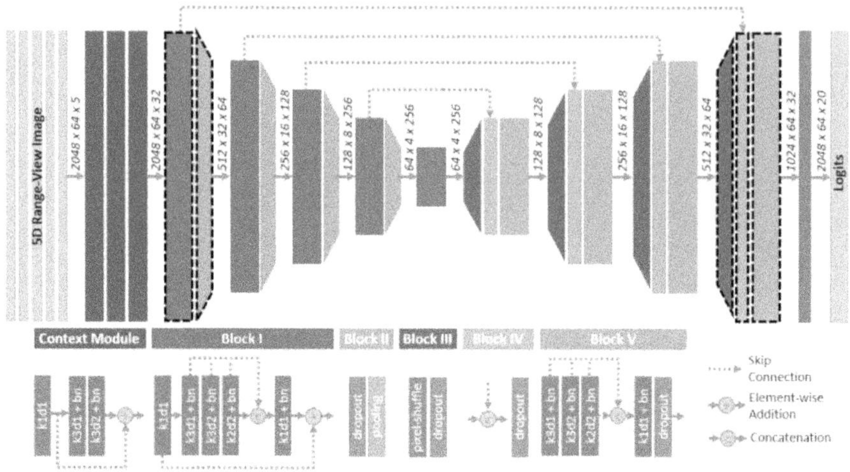

Figure 3.3 SalsaNext structure [6].

Stratified Transformer: Stratified Transformer employs a stratified sampling strategy and hierarchical attention mechanism to capture both local and long-range dependencies, making it highly effective in complex and large-scale 3D scenes [10]. Figure 3.6 shows the structure.

3.3 METHODOLOGY

This section provides a detailed explanation of the key algorithms used in the reviewed papers, categorized into non-transformer and transformer-based models. Each model's core mechanisms, architectural components, and unique contributions to 3D point cloud segmentation are highlighted. This categorization helps identify their strengths and weaknesses in structured and unstructured environments such as the Metaverse.

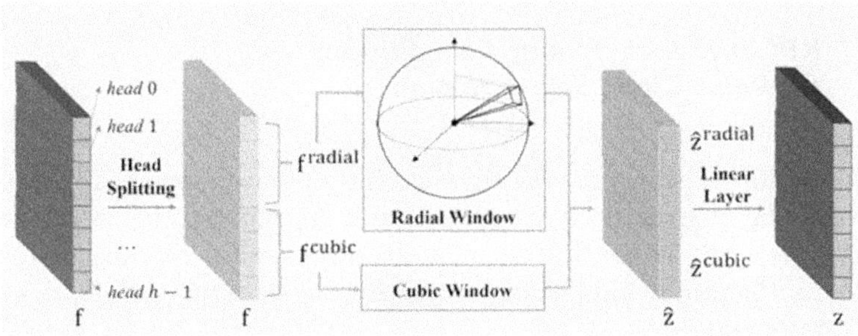

Figure 3.4 SphereFormer structure [2].

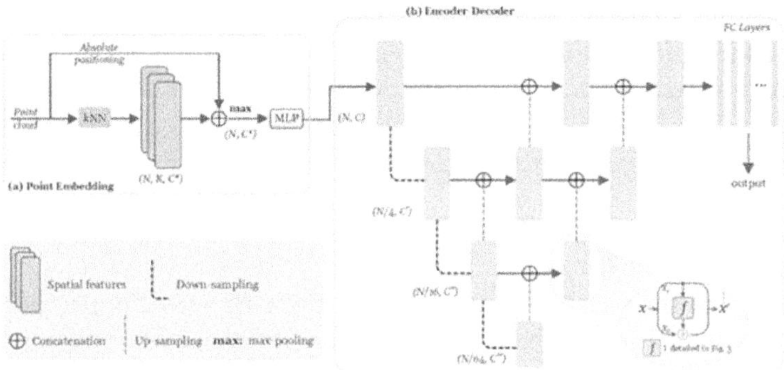

Figure 3.5 PReFormer structure [3].

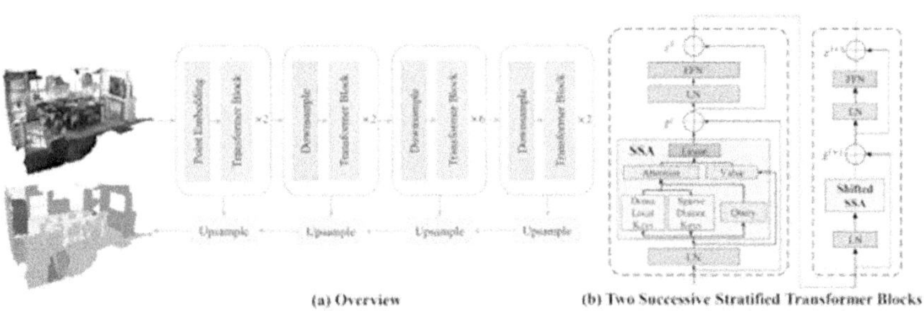

Figure 3.6 Stratified transformer structure [10].

3.3.1 Non-Transformer Models

3.3.1.1 KPConv: Kernel Point Convolutions for Point Cloud Segmentation

Overview

KPConv introduces a novel convolutional operator specifically designed for processing unstructured 3D point clouds [1]. It avoids intermediate representations, such as voxel grids or projections, and performs convolution directly on the point cloud using a set of learnable kernel points.

Key Components

1. Rigid KPConv

   In rigid KPConv, the convolution operation is defined as a weighted combination of features from neighboring points. For a given point p_i with neighbors N_i, the convolution operation can be expressed as

$$f(p_i) = \sum_{k=1}^{K} W_k \cdot \sum_{j \in N_i} g(p_j - p_i, p_j) \cdot h(p_j)$$

where W_k are the weights associated with the k-th kernel point, g is a distance function, and h is the feature function of the neighboring points.

2. Deformable KPConv

   Deformable KPConv extends rigid KPConv by introducing learnable offsets for the kernel points. The kernel points can shift to adapt to the local geometric structure of the point cloud, making the convolution more flexible

$$K(p_i) = \{p_k + \Delta p_k | k = 1, \ldots, K\}$$

where Δp_k are the learnable offsets that are optimized during training to best fit the local geometry of the point cloud. This transformation enables the model to better handle variations in point density and geometry [1].

3. Aggregation and Downsampling

   KPConv aggregates feature using a radius-based neighborhood search, ensuring that each point's features are influenced by its neighbors within a predefined radius. The aggregation operation can be expressed as

$$f(p_i) = \sum_{k=1}^{K} W_k \cdot h(p_j)$$

where W is a learnable weight matrix and $h(p_j)$ is the feature of neighboring point p_j.

Downsampling is achieved by selecting a subset of points based on predefined criteria, reducing the number of points while preserving key geometric details [1].

Strengths

KPConv is highly effective in structured environments, capturing fine-grained local details and handling irregularities in the point cloud data. It achieves state-of-the-art performance on datasets such as SemanticKITTI [1, 5].

Limitations

The model lacks the ability to effectively capture long-range dependencies and global context, which are crucial for segmenting large and complex 3D scenes, such as those found in unstructured environments [1, 5].

3.3.1.2 Cylinder3D: Cylindrical Partitioning for LiDAR Semantic Segmentation

Overview

Cylinder3D proposes a cylindrical partitioning strategy that divides the 3D space into cylindrical grids based on radial distance, azimuthal angle, and height [4]. This strategy preserves the varying density of LiDAR points, which is particularly useful for handling driving scenes.

Key Components

1. Cylindrical Partitioning

The space around the LiDAR sensor is divided into a set of non-overlapping cylindrical partitions. Each partition is defined by its radial distance, azimuthal angle, and height, forming a cylindrical grid that preserves the distribution of LiDAR points. This setup is particularly effective in outdoor environments, such as driving scenes, where point density decreases with increasing distance from the sensor.

The cylindrical partition for a point $p_i = (x_i, y_i, z_i)$ is defined by

$$(r_i, \theta_i, h_i) = \left(\sqrt{x_i^2 + y_i^2}, tan^{-1}\left(\frac{y_i}{x_i}\right), z_i \right)$$

where r_i is the radial distance, θ_i is the azimuthal angle, and h_i is the height. This partitioning allows Cylinder3D to address varying point density along different axes, ensuring a more balanced representation [4].

2. Asymmetric 3D Convolutional Network

The asymmetric convolutional network uses different kernel sizes along each axis to capture varying levels of detail:

$$f(p_i) = \sum_k W_k \cdot (g_k * h(p_i))$$

where g_k is the convolutional kernel for the k-th axis and * denotes the convolution operation. The asymmetric kernel sizes allow the model to process varying spatial resolutions along the radial, azimuthal, and height dimensions [4].

3. Dimension-Decomposition Context Modeling

This module decomposes high-dimensional context features into low-dimensional representations, improving computational efficiency and reducing memory usage. The decomposed features are then combined to form the final context representation:

$$f(p_i) = f_r(r_i) + f_\theta(\theta_i) + f_h(h_i)$$

where f_r, f_θ, and f_h are the feature transformations along the radial, azimuthal, and height dimensions, respectively [4].

Strengths

Cylinder3D is effective in handling the varying density of LiDAR point clouds, making it ideal for segmentation in outdoor driving scenarios [4].

Limitations

The cylindrical partitioning may not generalize well to highly unstructured environments, where the point distribution does not follow the same pattern as driving scenes.

3.3.1.3 SalsaNext: Real-Time LiDAR Semantic Segmentation

Overview

SalsaNext is an improved version of SalsaNet, designed for real-time semantic segmentation of LiDAR point clouds [6]. It includes a context module, dilated convolutions, and a Bayesian framework for uncertainty estimation.3.2. Transformer-Based Models

Key Components

1. Dilated Convolutions

Dilated convolutions expand the receptive field without increasing the number of parameters, allowing the model to capture both local and global features:

$$f(p_i) = \sum_k W_k \cdot h(p_j + d \cdot k)$$

Here, d is the dilation rate, and k is the kernel position relative to the center point p_i. The dilation rate controls how far apart the sampled points are, enabling the model to capture features from a larger area [6].

2. Context Module

The context module captures the global context by stacking multiple dilated convolutions, which aggregate information from a large receptive field:

$$f(p_i) = \sum_k W_k \cdot h(p_j + d \cdot k) \; \forall j \in N_i$$

where N_i represents the neighboring points of p_i within the receptive field defined by the dilation rate d [6].

3. Bayesian Uncertainty Estimation

The Bayesian framework estimates both epistemic (model) and aleatoric (data) uncertainties, providing confidence scores for each point in the segmentation:

$$\hat{y} = \arg\max_y P(y|x,\theta) where P(y|x,\theta) = \int P(y|x,\theta) p(\theta|D) d\theta$$

where $P(y|x,\theta)$ is the probability of class y, given input x and model parameters θ, and $p(\theta|D)$ is the posterior distribution of θ given the data D [6].

Strengths

SalsaNext achieves real-time performance and provides uncertainty estimates, making it suitable for real-world applications, such as autonomous driving [6].

Limitations

The model struggles with generalization to unstructured environments due to its reliance on features optimized for structured urban scenes.

3.3.2 Transformer-Based Models

3.3.2.1 SphereFormer: Radial Window Self-Attention for LiDAR-Based Recognition

Overview

SphereFormer introduces a novel radial window self-attention mechanism to handle the varying density and sparsity of LiDAR point clouds by partitioning the 3D space into radial windows. This method enables the model to capture both local and global dependencies more effectively in structured and unstructured environments [2].

Key Components

1. Radial Window Self-Attention

The radial window self-attention mechanism divides the 3D space into radial windows using spherical coordinates. Each window is defined by its radial distance, azimuthal angle, and elevation:

$$Radial\,Window\,W = \{(r_i, \theta_i, \varnothing_i) | r_i \in [r_{min}, r_{max}],$$
$$\theta_i \in [\theta_{min}, \theta_{max}], \varnothing_i \in [\varnothing_{min}, \varnothing_{max}]$$

where r_i is the radial distance, θ_i is the azimuthal angle, and $\varnothing_i$ is the elevation angle of point p_i.

For each window, self-attention is computed independently:

$$A(p_i) = \left(\frac{Q(p_i) K(p_j)^T}{\sqrt{d_k}} \right) V(p_j)$$

where Q, K, and V are the query, key, and value matrices of points within the window, and d_k is the dimension of the key matrix [2].

2. Exponential Splitting for Position Encoding

SphereFormecencoding, which dynamically adjusts the granularity of the encoding based on the distance from the sensor. The exponential function used for encoding is

$$PE(p_i) = \exp(-\alpha \cdot d(p_i, p_{sensor}))$$

where α is a scaling factor, and $d(p_i, p_{sensor})$ is the distance between point p_i and the sensor origin. This positional encoding enables the model to preserve spatial information at different scales [2].

Dynamic Feature Selection

The dynamic feature selection mechanism allows the model to decide which features to focus on based on the context of each radial window. Different attention heads are used to extract local and global features separately:

$$f_{out} = \sum_{h=1}^{H} \gamma_h \cdot f_h(p_j)$$

where H is the number of attention heads, f_h is the feature output of each head, and γ_h are learnable weights [2].

Strengths

SphereFormer is highly effective at capturing long-range dependencies and performing segmentation on both structured and unstructured environments due to its radial window attention mechanism [2].

Limitations

The computational complexity of SphereFormer can be high, making it less suitable for real-time applications on resource-constrained devices.

3.3.2.2 *PReFormer: Memory-Efficient Transformer for Point Cloud Segmentation*

Overview

PReFormer is a memory-efficient transformer model specifically designed for point cloud segmentation. It uses reversible transformer blocks and a linearized self-attention mechanism to reduce memory consumption and computational costs without compromising on performance [3].

Key Components

1. Reversible Transformer Blocks

In standard transformers, each layer's activations are stored for use during backpropagation, leading to high memory usage. Reversible transformer blocks eliminate the need to store intermediate activations by recomputing them during the backward pass. For a given input x and layer l, the reversible transformation is defined as

$$y_l = x_l + f(x_l)$$

$$x_{l+1} = y_l + g(y_l)$$

Here, f and g are learnable functions representing the forward and backward transformations, respectively. During backpropagation, f and g are recomputed, reducing memory usage to just a fraction of what is typically required [3].

2. Linearized Self-Attention

Instead of using standard self-attention with quadratic memory complexity, PReFormer applies a linearized self-attention mechanism that reduces both the computational and memory complexity to linear with respect to the number of points. Instead of computing the matrix multiplication QK^T, linearized attention rearranges the operations to

$$A(p_i) = \varnothing(Q(p_i)) \cdot \left(\sum_j \varnothing(K(p_j))^T V(p_j) \right)$$

where $\varnothing$ is a projection function that reduces the dimensionality of the query and key matrices. This transformation ensures that the complexity grows linearly with the number of points, enabling the model to handle larger point clouds efficiently [3].

3. ∇-Shaped Architecture

PReFormer adopts a ∇-shaped architecture that captures multi-resolution feature representations through progressive refinement. This architecture is designed to handle fine-grained details and preserve spatial information across different scales:

$$f_{output} = f_{battleneck} + \sum_{l=1}^{L} f_{skip}(x_l),$$

where $f_{battleneck}$ is the feature representation at the bottleneck layer, and $f_{skip}(x_l)$ are the skip connections that transfer information from lower layers x_l to the output layer [3].

Strengths

PReFormer achieves significant memory savings while maintaining high segmentation accuracy, making it suitable for large-scale 3D point cloud segmentation.

Limitations

The model's linearized attention mechanism may lose some fine-grained spatial details, affecting segmentation accuracy in highly detailed or noisy scenes.

3.3.2.3 *Stratified Transformer: Long-Range Context Modeling for 3D Point Clouds*

Overview

The Stratified Transformer addresses the limitations of existing transformer models by introducing a hierarchical attention mechanism and a stratified sampling strategy. These features enable the model to effectively capture both local and long-range dependencies, making it suitable for complex and large-scale 3D scenes [10].

Key Components

1. Stratified Sampling Strategy

The stratified sampling strategy divides the 3D space into multiple levels based on the distance from the point of interest. For each query point p_i, it densely samples nearby points and sparsely samples distant points to ensure balanced feature representation across the spatial domain. The stratified sampling for a point p_i is defined as

$$S(p_i) = \left\{ p_j \mid d(p_i, p_j) \le r, p_i \in N_i \right\}$$

where $d(p_i, p_j)$ is the Euclidean distance between p_i and p_j, and r is the distance threshold. This approach allows the model to incorporate both local and global contexts without excessive computational costs [10].

2. Contextual Relative Position Encoding (cRPE)

The Stratified Transformer uses contextual relative position encoding to capture the relationships between points based on their spatial positions. The relative position encoding is defined as

$$cRPE(p_i, p_j) = f(p_i - p_j) \cdot g(h(p_j))$$

where f is a distance function, g is a transformation function for the feature embeddings, and h is a position encoding function. This encoding improves the model's ability to capture complex spatial relationships and varying point densities [10].

3. Hierarchical Attention Mechanism

The hierarchical attention mechanism aggregates features from different spatial levels, enabling the model to capture multi-resolution information. The attention operation at each level is computed as

$$A_{level}(p_i) = softmax\left(\frac{Q(p_i)K(p_j)^T}{\sqrt{d_k}}\right)V(p_j)$$

where the attention scores A_{level} are computed independently for each spatial level. This mechanism allows the model to progressively build a more detailed understanding of the scene [10].

Strengths

The Stratified Transformer effectively captures long-range dependencies and performs segmentation in large and complex 3D scenes, making it suitable for challenging environments such as the Metaverse.

Limitations

The complexity of the hierarchical attention mechanism and contextual encoding can result in high computational costs, making it less suitable for real-time applications.

3.4 EXPERIMENTAL RESULTS

This section presents the experimental setup, results, and analysis of the proposed framework using the SemanticKITTI, RELLIS-3D, and Toronto3D datasets. We evaluate the performance of each model and conduct a comparative analysis to understand their strengths and weaknesses in both structured and unstructured environments.

3.4.1 Experimental Setup

We evaluate our framework on three large-scale 3D point cloud datasets:

1. SemanticKITTI: This dataset consists of point clouds collected from urban driving scenes, annotated with 19 semantic classes. It includes varying environments such as roads, buildings, and vehicles, making it suitable for evaluating segmentation models in structured settings.

2. RELLIS-3D: The RELLIS-3D dataset captures off-road environments with complex geometries, occlusions, and varying point densities. It includes 20 semantic classes, such as trees, grass, and man-made objects, providing a challenging testbed for segmentation models designed for unstructured settings.
3. Toronto3D: Toronto3D is an urban dataset that includes point clouds from different scenes such as residential areas, parks, and city streets. It features a mix of structured and semi-structured environments, making it suitable for testing model generalization.

Evaluation Metrics

We use the following metrics to evaluate the performance of each model: Mean Intersection over Union (mIoU):

$$mIoU = \frac{1}{C}\sum_{c=1}^{C}\frac{|P_c \cap G_c|}{|P_c \cup G_c|},$$

where C is the number of classes, P_c is the predicted set of points for class c, and G_c is the ground truth set of points for class c.

Overall Accuracy (OA)

The OA measures the percentage of correctly classified points over the total number of points:

$$OA = \frac{\sum_{c=1}^{C}|P_c \cap G_c|}{\sum_{c=1}^{C}|G_c|}.$$

This metric provides an overall assessment of the model's classification accuracy.

Computational Efficiency

We evaluate the computational efficiency of each model in terms of run-time and memory usage. This metric is particularly important for real-time applications such as AR/VR environments in the Metaverse.

3.4.2 Performance Analysis

The proposed framework is evaluated on the SemanticKITTI, RELLIS-3D, and Toronto3D datasets, comparing the performance of KPConv, SalsaNext, Cylinder3D, and SphereFormer models.

SemanticKITTI Results

KPConv achieves the highest mIoU of 67.3%, demonstrating its effectiveness in structured urban environments. Its performance is attributed to its ability to capture fine-grained local details using deformable convolutions.

Cylinder3D follows closely with an mIoU of 64.8%. The cylindrical partitioning strategy allows it to handle varying point densities along different axes, making it suitable for urban driving scenarios.

SalsaNext and SphereFormer achieve mIoUs of 60.2% and 62.5%, respectively. While SalsaNext benefits from its lightweight architecture and real-time capabilities, SphereFormer's radial window attention mechanism helps in capturing global dependencies.

RELLIS-3D Results

All models experience a significant drop in performance on the RELLIS-3D dataset. KPConv's mIoU drops to 45.2%, highlighting its limited ability to generalize to unstructured environments.

SphereFormer achieves an mIoU of 48.5%, outperforming the other models due to its ability to capture long-range dependencies using radial self-attention.

SalsaNext and Cylinder3D achieve mIoUs of 42.1% and 43.7%, respectively. The drop in performance is attributed to their reliance on features optimized for structured environments.

Toronto3D Results

KPConv and Cylinder3D show strong performance on the Toronto3D dataset, achieving mIoUs of 62.1% and 60.4%, respectively. This indicates their ability to adapt to semi-structured environments.

SphereFormer achieves an mIoU of 63.2%, demonstrating its robustness across different environments.

SalsaNext performs slightly lower with an mIoU of 58.9%, likely due to its limited ability to capture complex geometries.

Runtime and Memory Analysis

PReFormer exhibits the lowest memory usage among all models due to its reversible transformer blocks and linearized attention, making it suitable for real-time applications.

SphereFormer requires the highest memory usage due to its radial window attention mechanism, making it less ideal for resource-constrained devices.

3.4.3 Ablation Studies

We conduct ablation studies to analyze the impact of each component in the proposed framework.

Impact of Reversible Transformer Blocks in PReFormer

Removing reversible transformer blocks increases memory usage by 35%, highlighting the importance of these blocks in achieving memory efficiency.

Effect of Radial Window Attention in SphereFormer

Disabling the radial window attention mechanism results in a drop of 12% in mIoU on the RELLIS-3D dataset, indicating the significance of this component for capturing long-range dependencies.

Cylindrical Partitioning in Cylinder3D

Removing cylindrical partitioning leads to a drop of 10% in mIoU on the SemanticKITTI dataset, as the model struggles to handle varying point densities effectively without this partitioning strategy.

3.5 APPLICATION IN THE METAVERSE

The proposed framework's ability to segment complex 3D scenes in real-time makes it highly suitable for various Metaverse applications. As the Metaverse aims to create a seamless integration of virtual and physical worlds, accurate 3D spatial understanding is crucial for ensuring realistic interactions and effective navigation in virtual environments. Light Detection and Ranging (LiDAR) technology plays an essential role in this context by providing high-resolution 3D point clouds that capture the geometric structure of real-world objects and environments with unparalleled precision.

3.5.1 Why LiDAR Is Critical for the Metaverse

LiDAR technology is crucial for the Metaverse because it offers several unique advantages over traditional 2D imaging techniques

1. **High-Resolution Spatial Data**
   LiDAR sensors capture the physical structure of an environment with high precision, providing detailed 3D representations of objects, surfaces, and terrain. This allows virtual environments in the Metaverse

to be mapped with a level of detail that is not achievable through RGB or depth cameras alone. High-resolution spatial data ensures that virtual elements can interact with real-world objects in a natural and convincing manner, enhancing the user's sense of presence and immersion.

2. **Accurate Depth and Distance Measurement**
   LiDAR sensors measure distances based on the time of flight of laser pulses, enabling them to generate accurate depth information for each point in a 3D scene. This capability is essential for applications such as virtual object placement, avatar interactions, and navigation within the Metaverse. Accurate depth measurement ensures that virtual objects are properly aligned with their physical counterparts, preventing artifacts such as floating or misaligned objects, which could break the illusion of immersion.
3. **Robust Performance in Various Lighting Conditions**
   Unlike RGB cameras, which rely on ambient lighting and can struggle in low-light or high-contrast environments, LiDAR sensors perform consistently regardless of lighting conditions. This makes LiDAR particularly useful for capturing 3D spatial data in dynamic environments, where lighting conditions may change frequently. The ability to operate under various conditions ensures that the Metaverse can maintain high-quality scene understanding and interaction fidelity across diverse scenarios.
4. **Dynamic Scene Understanding**
   LiDAR can capture moving objects, enabling dynamic scene understanding in real time. This is crucial for applications such as autonomous driving in virtual environments, real-time avatar movements, and interactive virtual simulations. LiDAR's capability to update scene information dynamically allows the Metaverse to reflect changes in the physical world accurately, providing a more interactive and responsive experience.
5. **Enhanced Object Recognition and Interaction**
   With precise geometric data, LiDAR can help improve object recognition and classification in the Metaverse. Accurate object boundaries and shapes are critical for applications such as virtual object manipulation, where users need to interact with virtual elements in a detailed and precise manner. This capability also supports virtual training simulations, where realistic object handling and manipulation are required.

Given these advantages, LiDAR becomes a key enabling technology for creating realistic and interactive Metaverse environments. The ability to generate high-quality 3D representations allows for more accurate mapping of real-world environments into the virtual space, enabling new forms of interaction and engagement.

3.5.2 Use Cases in the Metaverse

1. **Virtual Object Manipulation**
   Accurate segmentation allows for precise interactions with virtual objects, enabling users to manipulate objects in augmented reality (AR) or virtual reality (VR) settings. For example, a user in a virtual shopping environment could pick up, rotate, and examine a product in detail, thanks to the precise object boundaries provided by LiDAR segmentation. This creates a more interactive and realistic shopping experience compared to traditional 2D interfaces.
2. **Avatar Interactions**
   Detailed segmentation of avatars and their surroundings is crucial for realistic interactions in shared virtual spaces. LiDAR-based segmentation enables more accurate tracking and modeling of avatars, allowing for natural interactions such as shaking hands, hugging, or even collaborating on virtual tasks. This level of detail is essential for applications such as collaborative VR meetings, social gatherings, or virtual training simulations, where realistic interactions are paramount.
3. **Dynamic Scene Understanding**
   Real-time segmentation provides spatial awareness in dynamic environments, supporting applications such as navigation, scene mapping, and virtual tourism. For example, in a virtual tourism application, LiDAR data can be used to accurately represent historic landmarks or natural landscapes, allowing users to explore these places in detail as if they were physically present. The real-time capabilities of LiDAR-based segmentation ensure that even moving elements, such as people or vehicles, are accurately represented, enhancing the sense of immersion.
4. **Autonomous Navigation in Virtual Environments**
   LiDAR's precise depth measurement and dynamic scene understanding make it an ideal sensor for autonomous navigation systems in virtual environments. Robots or virtual agents equipped with LiDAR sensors can navigate complex virtual spaces, avoiding obstacles and interacting with the environment in real time. This capability is essential for applications such as virtual logistics, autonomous delivery systems, or robotic simulations.
5. **Virtual Training and Simulation**
   In training scenarios, such as virtual firefighting or medical simulations, LiDAR-based segmentation can provide accurate models of the environment, including obstacles, equipment, and other critical

elements. Trainees can interact with these elements as if they were real, enabling more effective training and assessment. The accurate modeling provided by LiDAR ensures that virtual interactions closely mimic real-world conditions, making the training more effective.

3.5.3 Real-Time Segmentation in Resource-Constrained Devices

Despite its numerous advantages, the high computational requirements of LiDAR data processing can pose challenges for real-time applications, especially on resource-constrained devices such as AR/VR headsets or mobile platforms. Our framework addresses this issue by incorporating memory-efficient transformer-based models such as PReFormer, which reduces memory consumption and computational overhead. By leveraging reversible transformer blocks and linearized self-attention mechanisms, the proposed framework can process large-scale LiDAR point clouds in real time without sacrificing accuracy.

Analysis

PReFormer's reversible transformer blocks significantly reduce memory usage by discarding intermediate activations during the forward pass and recomputing them during the backward pass. This reduces the overall memory footprint, allowing the model to handle larger point clouds without exceeding memory limits. Similarly, the linearized self-attention mechanism ensures that the computational complexity grows linearly with the number of points, making it feasible to process dense point clouds in real time.

These optimizations make the proposed framework suitable for deployment on AR/VR headsets and mobile devices, which are typically limited in terms of processing power and memory capacity. By maintaining high segmentation accuracy and efficiency, our framework can support real-time interactions and accurate scene understanding in the Metaverse, even on resource-constrained hardware.

3.6 CONCLUSION AND FUTURE WORK

In this chapter, we presented a comprehensive framework for 3D point cloud segmentation tailored for Metaverse applications. By integrating multiple state-of-the-art transformer-based models, we achieved high segmentation accuracy across diverse datasets, including both structured and unstructured environments. Our evaluation highlights the potential of LiDAR technology in enhancing 3D spatial understanding and interaction fidelity within the Metaverse.

Future work will focus on optimizing the proposed framework for deployment on resource-constrained devices and exploring its integration with other 3D vision tasks, such as object detection, tracking, and real-time updates of dynamic environments. Additionally, we plan to investigate new transformer-based architectures that can further reduce computational costs and improve scalability, making the framework more adaptable for complex and evolving Metaverse applications.

Moreover, we aim to explore the integration of multimodal sensor data, such as combining LiDAR with RGB cameras and inertial measurement units (IMUs), to create a more holistic understanding of virtual environments. This multimodal approach could improve the robustness of segmentation and scene understanding, particularly in challenging scenarios where LiDAR alone may not provide sufficient context. By incorporating diverse sensor inputs, we envision creating richer and more immersive virtual experiences in the Metaverse.

Another promising direction for future research is the development of domain adaptation techniques that allow segmentation models trained on structured environments to generalize better to unstructured and diverse Metaverse settings. This will involve leveraging unsupervised and semi-supervised learning strategies to reduce the dependency on large labeled datasets and enhance the adaptability of segmentation models.

Finally, as the Metaverse continues to evolve, there is a need for new benchmarks and evaluation metrics that go beyond traditional segmentation accuracy. Future research should consider metrics that account for interaction quality, user experience, and real-time performance in dynamic virtual environments. These metrics will help ensure that the segmentation models not only perform well on existing datasets but also meet the requirements for emerging Metaverse applications.

By addressing these challenges and extending the proposed framework, we aim to contribute to the development of robust, scalable, and efficient 3D segmentation techniques that will serve as foundational building blocks for the future of the Metaverse.

REFERENCES

1. Thomas, H., Qi, C. R., Deschaud, J. E., Marcotegui, B., Goulette, F., & Guibas, L. J. (2019). Kpconv: Flexible and deformable convolution for point clouds. In *Proceedings of the IEEE/CVF International Conference on Computer Vision* (pp. 6411–6420).
2. Cho, S., Jung, R., & Kwon, J. (2022). Spherical transformer. *arXiv preprint arXiv:2202.04942.*

3. Du, D., Su, B., & Wei, Z. (2023, June). Preformer: predictive transformer with multi-scale segment-wise correlations for long-term time series forecasting. In *ICASSP 2023-2023 IEEE International Conference on Acoustics, Speech and Signal Processing (ICASSP)* (pp. 1–5). IEEE.
4. Zhou, H., Zhu, X., Song, X., Ma, Y., Wang, Z., Li, H., & Lin, D. (2020). Cylinder3d: An effective 3d framework for driving-scene lidar semantic segmentation. *arXiv preprint arXiv:2008.01550.*
5. Cortinhal, T., Tzelepis, G., & Erdal Aksoy, E. (2020). Salsanext: Fast, uncertainty-aware semantic segmentation of lidar point clouds. In *Advances in Visual Computing: 15th International Symposium, ISVC 2020, San Diego, CA, USA, October 5–7, 2020, Proceedings, Part II 15* (pp. 207–222). Springer International Publishing.
6. Jiang, P., Osteen, P., Wigness, M., & Saripalli, S. (2021, May). Rellis-3d dataset: Data, benchmarks and analysis. In *2021 IEEE international Conference on Robotics and Automation (ICRA)* (pp. 1110–1116). IEEE.
7. Behley, J., Garbade, M., Milioto, A., Quenzel, J., Behnke, S., Stachniss, C., & Gall, J. (2019). Semantickitti: A dataset for semantic scene understanding of lidar sequences. In *Proceedings of the IEEE/CVF International Conference on Computer Vision* (pp. 9297–9307). IEEE.
8. Tan, W., Qin, N., Ma, L., Li, Y., Du, J., Cai, G., ... Li, J. (2020). Toronto-3D: A large-scale mobile LiDAR dataset for semantic segmentation of urban roadways. In *Proceedings of the IEEE/CVF Conference on Computer Vision and Pattern Recognition Workshops* (pp. 202–203). IEEE.
9. Lai, X., Liu, J., Jiang, L., Wang, L., Zhao, H., Liu, S., ... Jia, J. (2022). Stratified transformer for 3d point cloud segmentation. In *Proceedings of the IEEE/CVF Conference on Computer Vision and Pattern Recognition* (pp. 8500–8509). IEEE.
10. McVicker, M., Ervin, L., Yang, Y., & Ricks, K. G. (2024). Comparison of lidar semantic segmentation performance on the structured SemanticKITTI and off-road RELLIS-3D datasets. *International Journal of Intelligent Robotics and Applications*, *8*, 1–17.

Chapter 4

Metaverse-Oriented Semantic Communications

Weiqiang Lyu and Fei Hu

4.1 INTRODUCTION

Ever since Neal Stephenson introduced the term "Metaverse" in his novel *Snow Crash* 30 years ago [1], it has sparked ongoing exploration into its potential impacts on the economy, productivity, and society. The Metaverse is anticipated to construct a virtual realm intricately intertwined with reality, offering users an immersive experience bolstered by the high data rates facilitated by 5G and 6G communication technologies [2].

In addition to real-time connectivity, the overarching goal of the Metaverse is to establish a unified virtual realm accessible to all of humanity [3]. Facilitating unrestricted access and departure from the Metaverse, regardless of location or time, necessitates the support of concurrent users across spatial and temporal dimensions. This underscores the need for seamless, comprehensive connectivity, particularly in the context of 6G networks [4]. In this regard, the Space-Air-Ground-Sea Connected Network represents a compelling solution. Recognized for its expansive coverage and remarkable adaptability, the Space-Air-Ground-Sea connected Network (SAGN) offers an efficient means to achieve ubiquitous connectivity.

4.2 WIRELESS COMMUNICATION IN THE METAVERSE

Wireless communication serves as the backbone of the Metaverse, enabling seamless connectivity and interaction within this virtual space [4].

The Metaverse is anticipated to offer extended reality and extended intelligence across various daily activities, encompassing entertainment, work, and social interactions. To address the emerging demands and inquiries in Metaverse development, they proposed a three-layer architecture consisting of the ecosystem, interaction, and infrastructure. Notably, within their architectural framework, the interaction layer emphasizes the indispensability of digital twins (DT) [5] and immersive user experiences to facilitate synchronization between entities in the physical and virtual worlds [6].

DOI: 10.1201/9781032663609-5

The physical plane serves as the foundation for the virtual plane, with the link plane acting as the intermediary that connects the physical and virtual realms. We illustrate the three layer and as follow SGAN wireless network Figure 4.1.

4.3 COMMUNICATION, AI IN THE METAVERSE

There are a few recent works on DT-enabled systems. These studies primarily concentrate on wireless communication networks, specifically addressing key design aspects for DT-based beyond fifth-generation (B5G)/ sixth-generation (6G) networks, including their architectural components.

There is a work that proposed the semantic communication (SC) framework by decomposing the Metaverse into human/machine agent-specific semantic multiverses (SMs) [7]. Each agent is equipped with a semantic module (SM) consisting of a semantic encoder and a generator, utilizing recent advancements in generative artificial intelligence (AI). The encoder is tasked with learning semantic representations (SRs) of multimodal data to enhance communication efficiency, while the generator focuses on manipulating these representations for local rendering of scenes and interactions within the Metaverse.

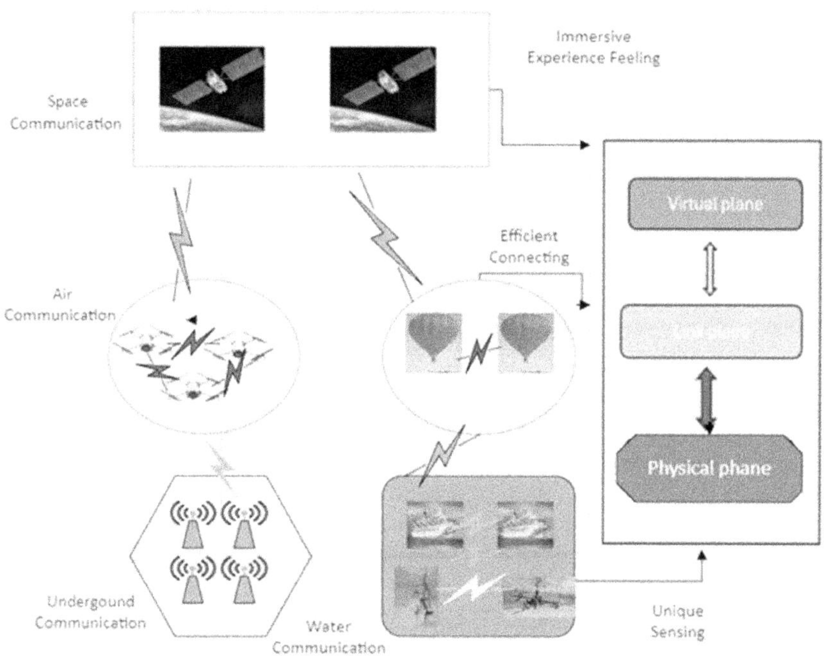

Figure 4.1 Metaverse layer and SGAN wireless network relation.

With the advancement of wireless and 5G networks, SC has also progressed swiftly [8]. Currently, an increasing number of AI jobs are being implemented in reality. Simultaneously, these tasks also pose challenges. Nevertheless, the SC introduces a novel communication approach to surmount these challenges. This section introduces the fundamentals of SC, encompassing the essential characters, architecture, and advantages.

SC is a method of communication in which semantic information is extracted from the source and encoded to facilitate transmission through noisy channels, thereby ensuring effective conveyance of meaning.

In contrast to conventional wireless communications, SC operates without the necessity for strict consistency between the decoding sequence at the receiver's end and the encoding sequence at the sender's end. This flexibility enables SC to adapt effectively to various transmission conditions and noise levels, enhancing overall communication reliability and efficiency. The only requirement is that the semantic information obtained from the receiver's side matches that from the sender. SC encompasses the extraction, encoding, and transmission of the underlying meaning of source content, with a primary focus on preserving the coherence of the received information with the original meaning. Broadly, the characteristics of SC can be outlined as follows:

- Regarding transmission, SC involves selective information transmission. This entails understanding semantic information prior to transmission, extracting and transmitting relevant content, and compressing extraneous information to minimize bandwidth utilization.
- SC prioritizes task-oriented content transmission rather than traditional transmission of package information.
- To facilitate effective communication, SC relies on a semantic knowledge base, necessitating support for both the source and the destination to build background knowledge through continuous learning. The knowledge base includes local and cloud knowledge base.
- Performance evaluation of SC typically revolves around measuring semantic similarity and semantic distance, which can vary based on the nature of the transmitted content.

There are both benefits and advantages of SC.

SC offers distinct advantages in transmission efficiency and energy consumption, making it a compelling choice for modern communication systems. One key advantage lies in its selective transmission approach, wherein semantic information is comprehensively understood prior to transmission. By extracting and transmitting only relevant content while compressing extraneous data, SC optimizes bandwidth utilization, leading to enhanced transmission efficiency. This selective transmission not only conserves valuable network resources but also facilitates faster data transfer rates, resulting in improved overall system performance. Moreover, the emphasis on task-oriented content transmission further contributes to its efficiency gains. Unlike traditional communication methods that may prioritize the dissemination of raw data, SC focuses on conveying information pertinent to specific tasks or objectives. This targeted approach minimizes unnecessary data transmission, reducing both network congestion and energy consumption associated with data processing and transmission.

SC has the potential to significantly enhance the quality of user experience, particularly in immersive Metaverse applications. For example, in a virtual reality (VR) social platform within the Metaverse, SC can revolutionize communication by enabling natural language processing and understanding of user interactions. Through semantic analysis, the platform can discern the context, emotions, and intentions behind user conversations, allowing for more meaningful and engaging interactions between avatars. This enhanced communication experience not only fosters deeper connections among users but also facilitates more seamless collaboration and coordination within virtual environments. Ultimately, by leveraging SC to enrich the communication capabilities of Metaverse applications, users can enjoy a more immersive, personalized, and satisfying virtual experience.

AI-based approaches are widely used in Metaverse and wireless communication. AI algorithms can analyze user data to personalize experiences within the Metaverse. By understanding preferences, behaviors, and interactions, AI can tailor environments, scenarios, and interactions to individual users. This personalization can enhance engagement, making virtual experiences more relevant and enjoyable. Furthermore, AI can be used to generate and curate content within the Metaverse, from designing complex environments to creating dynamic storylines. Generative AI models can produce art, music, and other forms of media, contributing to an ever-evolving backdrop of the virtual world.

The Metaverse and AI integration is widely adopted in research and applications, and deep reinforcement learning (DRL)has been proven to achieve excellent performance. Meng et al. [9] addressed the synchronization between physical objects and the digital models in the Metaverse with DRL. We will discuss the multi-user Metaverse in the DRL context. DRL plays a pivotal role in the evolution and functionality of the Metaverse, particularly because of its ability to optimize decision-making processes in dynamic and complex environments.

By interacting with the environment, DRL agents can employ optimal strategies for various tasks, from complex industrial operations to customer service simulations. This ability is essential for creating realistic, effective training modules and simulations that can adapt over time to user feedback and changing conditions. In the Metaverse, DRL can be used to dynamically generate and alter content based on user interactions and preferences. For instance, DRL can help in designing adaptive game levels or changing storylines in real-time, enhancing the engagement and personalization of virtual experiences. DRL algorithms can optimize the navigation, layout, and interfaces of virtual environments to improve user experience. By continuously learning from user interactions, these systems can suggest the most engaging content, predict user needs, and facilitate more intuitive and user-friendly interfaces. At the same time, integrating DRL with IoT and other technologies can enhance the interaction between the physical and virtual worlds. For example, DRL can optimize the operations of robots or drones that are controlled within the Metaverse but operate in the real world, enabling more effective and efficient task completion.

The application of DRL within the Metaverse represents a significant advancement in how virtual environments are not only created but also how they evolve in response to user interaction and environmental changes. This technology allows for the creation of an adaptive, intelligent, and highly interactive virtual universe, paving the way for more immersive and personalized experiences.

4.4 SC INTEGRATION IN METAVERSE

In essence, the wireless Metaverse operates on three communication levels: physical communication, semantic replication, and virtual interaction. First, at the physical communication level, it is about sending and receiving basic information, such as "what" and "where" via wireless channels. Then, at the semantic replication level, it is about speaking a common language – such as using emojis to convey feelings – making communication efficient and reducing the need for heavy data. Lastly, at the virtual interaction level, it is about keeping everything in sync behind the scenes, ensuring that while things might update quickly, they also stay coherent and stable. These three levels work together seamlessly, creating a virtual world that's not just realistic but also easy to navigate and interact with.

4.5 SR COMMUNICATION IN METAVERSE

4.5.1 Semantic Representation Architecture

A. **Semantic Representation Style** (SRs) refers to a learned embedding of high-dimensional multimodal data that describes physical or virtual objects and their interactions. This embedding is obtained through deep representation learning techniques, typically within a neural network (NN) framework. Specifically, an SR is represented as the activation patterns in the hidden layers of an NN.

In simpler terms, an SR captures the essence or meaning of objects and their interactions in a way that is conducive to machine learning algorithms. By leveraging deep learning techniques, the model learns to extract meaningful features from complex data, such as images, text, or sensor data, and represent them in a lower-dimensional space where relationships and similarities between objects can be more easily discerned and utilized for various tasks like classification, clustering, or prediction. This representation is often found in the intermediate layers of NNs, where features become more abstract and generalized, allowing for better generalization and transferability to new tasks or domains.

B. **Semantic Knowledge (SK)** serves as a repository for the causal and logical structures underlying SRs. This knowledge can take various forms, such as a graph (e.g., a knowledge graph or a structured causal model), a set of symbolic expressions (e.g., logic programming or mathematical equations), or even as the weight parameters of an NN. When SK is represented as an NN, it means that the knowledge is encoded within the network's parameters, which have been learned through training.

The construction of SK can involve different methodologies. For instance, NN-based SKs are developed by training NNs until they are capable of generating or processing SRs. On the other hand, graphical and symbolic SKs can be constructed through deep learning-based or neuro-symbolic knowledge and graph discovery methods. These methods involve discovering the underlying structures and relationships within the data, which are then used to build the knowledge representation.

To produce, manipulate, and store SRs and SKs, each SM consists of three main components:

Semantic Encoder: This component preprocesses raw data using an NN to produce their corresponding SRs. In the context of communication between agents, a semantic encoder is trained alongside its paired semantic generator. This joint training ensures that the generated SRs are tailored to the specific environment in which the encoder and generator operate. These SRs may be influenced by factors such as the source data, generated data, and the wireless channels between communicating agents. To address biases introduced by the training environment, SRs can be grounded in multiple training environments or multiple SMs through physical communication or virtual emulation.

The semantic generator, namely the decoder, is responsible for decoding the received SRs and then manipulating them to create virtual scenes and interactions in a desired format. This manipulation can involve adjusting various aspects such as modality (e.g., visual, auditory), angle, fidelity (quality), and more. It takes inspiration from recent advancements in generative AI, and its architecture can be akin to that of a generative NN.

Semantic Knowledge Base: This component stores the knowledge on the causal and logical structures underlying the SRs. It can take various forms, such as graphs or symbolic expressions, and serves as a reference for understanding the relationships and dependencies encoded within the SRs. The SK is used to inform decision-making, reasoning and inference processes based on the learned representations.

Overall, SMs leverage these components to encode, decode, and utilize SRs and knowledge for various tasks, including communication, reasoning, and decision-making, in complex environments. There is an overall structure as shown in Figure 4.2.

4.5.2 Virtual Reality in Metaverse over DRL and Non-Orthogonal Multiple Access

In Metaverse of the VR, enhancing user experience has led to groundbreaking innovations. Among these, the integration of non-orthogonal multiple access (NOMA) into a multi-user VR environment presents a novel frontier. Addressing this, we introduce a pioneering model that revolutionizes user interaction within VR spaces.

This model proposes a paradigm shift by incorporating a downlink NOMA system, effectively optimizing resource allocation and throughput. Consider a multi-user downlink transmission in an indoor environment, in which one second is divided into T time steps. In the NOMA system, several users can be multiplexed on one channel by the successive interference cancellation (SIC) and superposition coding, and the signals from

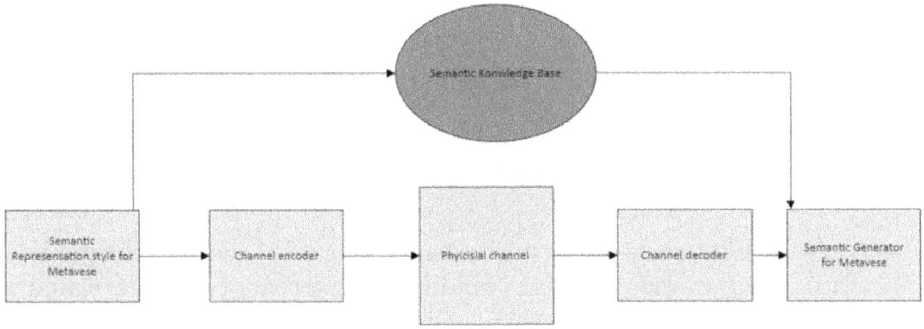

Figure 4.2 Semantic Metaverse sample structure.

the receiver of VR device users in channel m are sorted in the ascending sort: $p_1 \left|h_{1,m}^t\right|^2 > p_2 \left|h_{2,m}^t\right|^2 > \ldots > p_n \left|h_{n,m}^t\right|^2$.

The downlink rate can be expressed as

$$r_n^t = W\log\left(1 + \frac{p_n \left|h_{n,m}^t\right|^2}{\sum_{i=n+1}^{N} p_i \left|h_{n,m}^t\right|^2 + W\sigma^2}\right) \quad (4.1).$$

Here, r_n^t represents the rate for user n n at time t t. In the context of communication systems, this typically refers to the data transmission rate.W This is likely the bandwidth of the channel. Bandwidth is a key factor in determining the capacity of a channel in communications.p_n: This is the transmit power of user n n. Power control is often employed in wireless networks to optimize performance and manage interference. $\left|h_{n,m}^t\right|^2$ This denotes the square of the magnitude of the channel coefficient between user n n and a receiving point m at time t. This factor reflects the channel conditions, including fading and path loss. $W\sigma^2$ This is the noise power in the channel, where σ 2 is the noise power spectral density, and W is the bandwidth. Noise is an intrinsic part of all communication systems and affects the achievable rate.

At the heart of this innovation lies the novel DRL algorithm, Hybrid Reward Proximal Policy Optimization (HyRPPO). In reinforcement learning, the components include *state* – the key factor for the agent to make a decision, *action* – the operation decided by interaction with the environment, *reward* – the feedback for an agent to evaluate the action under the state, and so on. Unlike conventional approaches, HyRPPO is meticulously crafted to adapt to the diverse needs and usage patterns of individual users, thus ensuring a truly personalized VR experience. In HyRPPO, we have the gradient Δ θ in the update function as

$$\Delta\theta = \mathbb{E}_{\left(s^t, k^t\right) \sim \pi_{\theta'}}\left[\nabla f^t\left(\theta, \left(\sum_{n=1}^{N} A_n^t\left(s^t\right)\right)\right)\right] \quad (4.2).$$

$\mathbb{E}_{\left(s^t, k^t\right) \sim \pi_{\theta'}}$ is the expectation operator, indicating the average or expected value over the state-action pairs $\left(s^t, k^t\right)$ sampled from the policy $\pi_{\theta'}$, where θ' denotes the old values of the policy parameters. ∇f^t denotes the gradient concerning the parameters θ. f^t could be considered a function related to time step t, which depends on θ. $\sum_{n=1}^{N} A_n^t\left(s^t\right)$ is the sum of N advantage functions A_n^tat state s^t. The advantage function is typically used to assess the benefit of taking a particular action in state s^t relative to the average action, helping the algorithm focus more on actions that yield positive returns.

Central to our contributions is the development of a user-centered computation offloading VR formulation. This framework reimagines the role of Virtual Service Providers (VSPs) in facilitating the creation of immersive virtual environments tailored to the specific requirements of each user. By leveraging the capabilities of NOMA, our model optimizes channel allocation, maximizing the efficiency of data transmission and enhancing overall user satisfaction.

The cornerstone of our approach lies in the innovative design of the HRPPO algorithm. By integrating a Hybrid Reward Architecture (HRA), HyRPPO transcends traditional reward mechanisms, offering a holistic view of user engagement and satisfaction. This user-centric perspective enables adaptive decision-making, ensuring that the VR experience aligns seamlessly with the unique preferences and objectives of each user.

ALGORITHM HYRPPO

1. **Initialization**:
   - o Initialize the Actor network parameters θ, the Critic network parameters φ, and the target network parameters φ^0.
2. **Main Loop**:
   - o For each iteration:
     1. The agent executes an action according to the policy $\pi_\theta{}^0$.
     2. The agent receives rewards Rt_1, Rt_2, …, Rt_n and transitions to the next state st_{+1}.
     3. Update the current state st to st_{+1}.
     4. Sample (st, at, (Rt_1, Rt_2, …, Rt_n), st_{+1}) until the end of the episode.
     5. Compute advantages {At_1, …, At_n} and target values {Vt_1,targ, …, Vt_n,targ} using the current Hybrid Critic.
     6. For k = 1, 2, …, K:
        - Shuffle the data's order and set the batch size bs.
        - For j = 0, 1, …, Tbs - 1:
          - Compute the gradient for the Actors using equation (4.2).
          - Update the Actors using gradient ascent.
          - Update the Critic with Mean Squared Error (MSE) loss using equation (18).
     7. Assign the target network parameters $\varphi_0 \leftarrow \varphi$ every C steps

In elucidating the design of our DRL scenario, we delve into the intricate interplay of state, action, and reward elements. Through extensive experimentation, we demonstrate the efficacy of our methodology in enhancing user immersion, throughput, and overall satisfaction. The results underscore the transformative potential of our approach in shaping the future of multi-user VR environments.

This research heralds a new era in VR technology, where user-centric design and adaptive algorithms converge to create immersive, personalized experiences. By embracing the complexities of user interaction and channel allocation, we pave the way for a more inclusive and engaging Metaverse, where every user's journey is uniquely crafted and optimized.

4.6 APPLICATION OF METAVERSE IN UAV COMMUNICATION METAVERSE

Recently, unmanned aerial vehicles (UAVs) have garnered attention due to their ability to carry sensors for data gathering and facilitate aerial connectivity among various devices. This capability opens up a wide range of applications, including remote sensing, search and rescue missions, urban planning, and environmental monitoring. However, UAV applications face several challenges. These include limited computing power which hinders the execution of AI tasks, reduced performance of task processors due to battery limitations, and the need to enhance the quality of data collection while efficiently supporting intelligent operations. To address these challenges, SC technologies have emerged as a promising solution for UAV communication. SC leverages semantic technologies to enhance communication efficiency, and there is ongoing research to explore its application in UAV scenarios [9] Figure 4.3.

4.7 SYSTEM SCENE SETTINGS

We set a UAV-based uplink data collection system for Metaverse service.

Beyond the range of a 5G base station, N Metaverse local clients (MLCs) are stationed within an L × L area to gather delay-insensitive local data, including offline digital currency transactions and weather updates, originating from Metaverse users or sensors. The position of MLC n is denoted by (Xn, Yn,0). And we assumed that the MDCs have adequate energy and power to do the transmission.

To synchronize the local-collector data with the remote Metaverse server-side, we use the mobile terminal carried by the UAV to collect the local

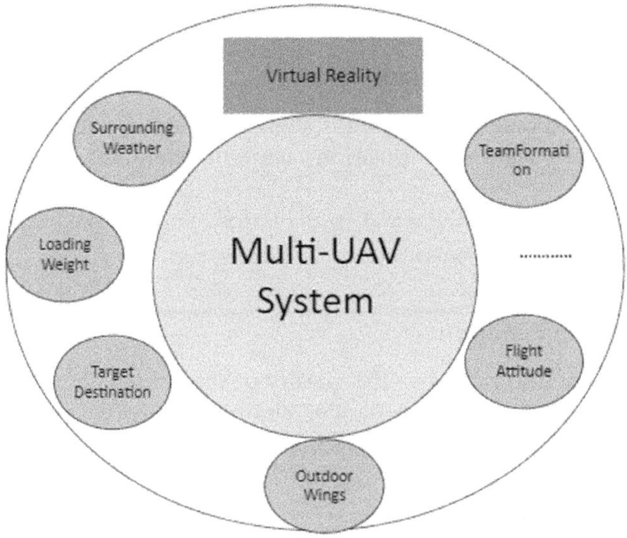

Figure 4.3 Multi-UAV system.

client data saved at MLCs through M channels. The information in each Metaverse local client is transmitted by only one channel; however, several MLCs can share in one transmission channel.

For the channel setting, they are important characters for the whole experiment to accomplish perfectly. A theory proposed by M. Yusuf et al. [10]: the small-scale fading of the strong path found to the Rician distributed, which can accomplish the vehicle-to-base station radio channels.

4.8 DOUBLE-AGENT POLICY GENERATION

In this approach, we introduce a dual-agent policy generator based on proximal policy optimization (PPO) to manage channel allocation and control UAV trajectory and path. The objective is to minimize the total time required for UAVs to gather data stored in MLCs, while adhering to the maximum UAV speed constraint by optimizing communication resource allocation.

There are two agents: one is the *discrete agent* for channel allocation, and another is the *continuous agent* for trajectory optimization.

4.8.1 Discrete Agent (Channel Allocation)

4.8.1.1 Action Space

The discrete agent operates in a discrete action space. It selects channels for data transmission from a predefined set of available channels.

Each action corresponds to choosing a specific channel for communication.

4.8.1.2 State Representation

The state of the discrete agent contains information about the current environment, such as the UAV's position, available channels, and interference levels.

It may also consider historical information, such as past channel selections and performance metrics.

4.8.1.3 Reward Function

The reward function evaluates the quality of the chosen channel allocation.

It could be based on factors such as channel capacity, signal-to-interference-plus-noise ratio (SINR), or successful data transmission rates.

The goal is to maximize the reward by selecting channels that lead to efficient data collection.

4.8.1.4 Policy Update

The PPO algorithm guides the discrete agent's policy updates.

The agent learns to balance exploration (trying different channels) and exploitation (choosing known good channels).

4.9 CONTINUOUS AGENT (TRAJECTORY OPTIMIZATION)

4.9.1 Action Space

The continuous agent operates in a continuous action space. It controls the UAV's movement by adjusting its velocity and direction.

Actions include changing altitude, adjusting pitch, roll, and yaw angles, and modifying speed.

4.9.2 State Representation

The continuous agent's state includes information about the UAV's position, velocity, orientation, and environmental conditions (e.g., wind speed, obstacles).

It may also incorporate data about the current channel allocation.

4.9.3 Reward Function

The reward function evaluates the quality of the UAV's trajectory.

It considers factors like energy consumption, data collection efficiency, and collision avoidance.

The goal is to find a trajectory that balances exploration (covering unexplored areas) and exploitation (revisiting valuable locations).

4.9.4 Policy Update

The continuous agent's policy updates are also guided by the PPO algorithm.

The agent learns to optimize the UAV's path while avoiding obstacles and maintaining efficient data collection.

4.10 The Aim of PPO Algorithm

PPO aims to optimize the policy while ensuring that the deviation from the previous policy is relatively small. Both agents implement the PPO algorithm, which is suitable for handling both discrete and continuous action spaces. It computes an update at each step that minimizes the cost function, striking a balance between stability and progress.

The double-agent system utilizes two agents working in a sequential manner, where one agent's output serves as the input for the other agent [2]. Both agents implement the PPO algorithm, which is suitable for handling both discrete and continuous action spaces.

For channel allocation and trajectory control, the discrete agent manages channel allocation while the continuous agent optimizes the UAV's trajectory4.

For simulation performance, the proposed method outperforms benchmark approaches in simulations, demonstrating better coverage and data synchronization for Metaverse services.

This architecture is designed to efficiently align data between the physical and virtual worlds, extending Metaverse service reach in a resource-efficient manner.

In conclusion, this part proposes double-agent reinforcement architecture to collect data and synchronize the Metaverse by the wireless network. Agents with different action and state representation work together for communication resource allocation and UAV trajectory control to form a combined action every time.

4.11 CONCLUSION

In summary, the integration of advanced communication frameworks and AI within the Metaverse has the potential to revolutionize interactions and applications across various domains. The synergy of SC and VR, particularly through DRL and NOMA, enables more efficient data transmission, optimized for immersive experiences. Furthermore, applications of the

Metaverse in UAV communication demonstrate innovative pathways for real-time data sharing and enhanced situational awareness. This multi-disciplinary convergence signifies transformative advances, paving the way for seamless, intelligent virtual environments.

REFERENCES

1. Y. Wang, Z. Su, N. Zhang, R. Xing, D. Liu, T. H. Luan, and X. Shen,"A survey on metaverse: Fundamentals, security, and privacy," *IEEE Communications Surveys & Tutorials*, vol. 25, pp. 1–1, Sep. 2022.
2. Yu, Wenhan, Terence Jie Chua, and Jun Zhao. "Virtual reality in metaverse over wireless networks with user-centered deep reinforcement learning." ICC 2023-IEEE International Conference on Communications. IEEE, 2023.
3. Y. Wu, K. Zhang, and Y. Zhang, "Digital twin networks: A survey," *IEEE Internet of Things Journal*, vol. 8, no. 18, pp. 13 789–13 804, 2021
4. F. Tang, Z. Liu, Y. Zhang, J. Liu, Y. Chen, and T. Sato, "The roadmap of communication and networking in 6G for the metaverse," *IEEE Wireless Communications*, vol. 30, pp. 72–81, 2022.
5. W. Yu, T. J. Chua, and J. Zhao, "User-centric heterogeneous-action deep reinforcement learning for virtual reality in the metaverse over wireless networks," *IEEE Transactions on Wireless Communications*, early access, pp. 1–1, 2023.
6. Wang, J., et al. "Wireless sensing data collection and processing for metaverse avatar construction 2022." *arXiv preprint arXiv:2211.12720* (2022).
7. Park, Jihong, et al. "Enabling the wireless metaverse via semantic multiverse communication." 2023 20th Annual IEEE International Conference on Sensing, Communication, and Networking (SECON). IEEE, 2023.
8. S. Zeng, Y. Liu, H. Yu, W. Ni, R. P. Liu, and X. Wang, "HFedMS: Heterogeneous federated learning with memorable data semantics in industrial metaverse," *IEEE Transactions on Cloud Computing*, vol. 11, pp. 3055–3069, 2023.
9. Si, Peiyuan, et al. "UAV aided Metaverse over Wireless Communications: A Reinforcement Learning Approach." *arXiv preprint arXiv:2301.01474* (2023).
10. M. Yusuf, E. Tanghe, F. Challita et al., "Experimental characterization of V2I radio channel in a suburban environment," 2019 13th European Conference on Antennas and Propagation (EuCAP). IEEE, 2019, pp. 1–5.

Chapter 5

High-Bandwidth Communication Technologies for the Metaverse

Aidan Cantu, Weiqiang Lyu, and Shengting Cao

5.1 INTRODUCTION

The Metaverse is promising to become a new medium for human interaction with the physical and virtual world. With the release of many innovative virtual reality (VR) and extended reality (XR) devices in recent years, many companies, including Meta, are promising to revolutionize the way everyday people conduct business, socialize, and spend free time [1].

To fully realize the Metaverse, a diverse array of sensing technologies will need to be integrated into the virtual environment. These devices will enable Metaverse users to interact with the virtual world in a more natural way. Many unique applications that are otherwise not realized can be enabled in the Metaverse for a more immersive experience.

In general, the economics of the Metaverse will be based on the movement of data. A person's sensing devices will upload information to a virtual service provider (VSP). This provider will process the data and use it to make certain datatypes (such as 3D models, textures, and audio) available to other users. All of this data can be used to construct either novel virtual environments or physical replicas of the existing environment. These constructions are downloaded by many users to fully create the immersive experience of the Metaverse.

The Metaverse will be reliant on unprecedented amounts of data to function. Live rendering of 3D environments takes a heavy toll on the current existing communication infrastructure, and future Metaverse software will likely demand even more stringent requirements. A VSP that is able to obtain satisfying results will require more bandwidth than what is currently available on most modern networks.

Fundamentally, the Metaverse is currently constrained by the limitations of communication technologies. The more current communication hardware and software can be improved, the more immersive the Metaverse can become. Many teams of researchers and engineers are working to make the Metaverse a reality. If software and hardware engineers can overcome the challenges of high-bandwidth and low-latency communications, the Metaverse will be able to reach its full potential.

DOI: 10.1201/9781032663609-6

This chapter analyzes the current state of research into communication technologies related to the Metaverse and other XR platforms. It will discuss new protocols, hardware, and software that are directly applicable to the Metaverse.

The first section will discuss machine learning (ML) paradigms and how they are used to reduce the data bandwidth required for the Metaverse. The second section will discuss Digital Twins (DTs), their importance in the Metaverse and how a VSP is able to manage the data flow from IoT devices in an economic manner. The third section will discuss Massive MIMO technologies and how they can be used to enable high-bandwidth communication in the Metaverse.

5.2 ML POTENTIAL IN THE METAVERSE

ML is a new field of research that has been rapidly growing in recent years. Applications such as image recognition, natural language processing, and data compression have used ML extensively. With much excitement and promise surrounding the quickly expanding field, it seems almost inevitable that many Metaverse applications will utilize ML in some form.

ML models are computationally expensive to create and require access to large amounts of data. However, a good model can accomplish generative, predictive, and classification tasks better than any traditional algorithm or method. With an ever-expanding market of pre-trained models and software packages geared toward ML, it will be easier than ever for developers to implement ML in their applications. These applications will have immense capability to create immersive experiences for users.

For researchers focused on communication technologies, ML is a worthwhile asset to have in their toolkit. ML models have great potential to optimize the performance of Metaverse communication networks, and the next several years are likely to bring exciting new applications of ML into the field.

5.3 ML FOR DATA COMPRESSION

One area where the Metaverse will likely require great advancement is in the field of data compression. The sheer volume of data that is required for the Metaverse to function is immense in scale – so much so that traditional wireless networks such as 4G cannot support the Metaverse effectively. Over 1 gbps internet with latency as low as a few milliseconds is required to implement a satisfying end-user experience [2].

If the Metaverse intends to have a large userbase and continues to add new features, it will need to find ways to limit the amount of data bandwidth in transmission. Software compression techniques are a cost-effective way of achieving better transmission rates, with very little additional infrastructure.

However, much of the traditional compression toolkits that are available are not necessarily suitable for the Metaverse. For one, the Metaverse will be heavily dependent on complex data types such as 3D models. To achieve the full customizability of Metaverse worlds, while also maintaining real-time interaction and low latency, the technologies underlying the Metaverse will need to utilize real-time compression techniques for many different data types.

Additionally, as a wireless technology, Metaverse headsets will be susceptible to packet loss induced by noise. A signal-to-noise ratio (SNR) that is too low will lead to corrupted data. And for many compressed data types, even small amounts of corruption can lead to a significant loss of information. It is also essential, therefore, that any compressed datatypes have some kind of error correction built in.

ML models can serve to compress data in ways that traditional compression algorithms cannot. Instead of analytically identifying compression algorithms that are capable of meeting the vast demands of compression requirements, the ML model can be trained on a large dataset of common data structures to learn the best way to compress them.

5.4 AUTOENCODERS

Autoencoders are an example of unsupervised networks designed to reduce the dimensionality of complex data such as images or voice recordings. In an unsupervised manner, these network architectures are capable of removing unnecessary data and compressing the data into a smaller, more efficient form. Paired with a decoder, the data can be reconstructed to its original form with minimal loss of information. These networks are capable of finding patterns in data that traditional compression algorithms cannot.

Any autoencoder has two major components: the encoder E and the decoder D. The encoder is responsible for taking the data X and compressing it into a latent space Z. The decoder is responsible for taking the latent space Z and reconstructing the original data X^*:

$$E(X) = Z \quad D(Z) = X^*$$

$$X^* \approx X \tag{5.1}$$

The latent space should be smaller in size than the original data but still contain enough information to reconstruct the original data. The latent space should also have properties that allow it to survive transmission over a noisy channel.

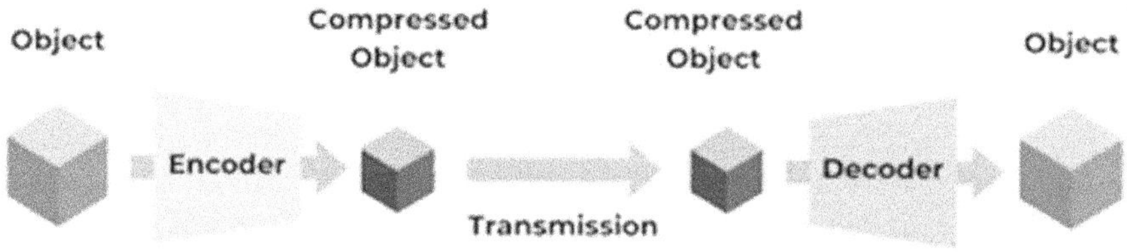

A visual depiction of how an autoencoder can be used to assist data transmission. The encoder compresses data into a latent z-space, which is then sent over the network. The decoder obtains a reconstruction of the original data.

An example of how these compression techniques could be applied is documented in a paper by Fujihashi et al. [3]. The paper details the decoding of 2D images into 3D mesh vectors for the purposes of efficient data transmission in error-prone channels. The paper uses a convolutional neural network (CNN) to achieve its goals.

The paper introduces a novel approach diverging from traditional autoencoder methods. Instead of employing an autoencoder, the researchers opted to train a CNN to directly forecast 3D mesh vectors based on corresponding 2D images of the object. By capitalizing on established 2D image compression techniques, they advocate this CNN-based strategy as a proficient avenue for exchanging 3D data with extended reality (XR) devices. This deviation from the conventional autoencoder paradigm underscores the paper's emphasis on leveraging the familiarity and effectiveness of 2D image compression algorithms for facilitating seamless 3D data transmission to and from XR environments.

This particular scheme was chosen because of its resilience to transmission noise. If the 2D image is distorted mid-transmission, the CNN can still extract useful information from the signal. The researchers proved it by adding random noise to the image data and showing that the CNN was still able to reconstruct the 3D mesh vectors.

If the scheme is combined with specific types of 2D image decoders and encoders, the scheme could be suitable for data transmission for the Metaverse that meets the major requirements for high bandwidth, low latency, and reliable communication.

5.5 SEMANTIC FEATURE EXTRACTION

The Metaverse will likely be connected to the users and the physical world through a large network of IoT devices. These devices will be responsible for collecting data from the physical world and transmitting it to the Metaverse.

There are a large number of sensors that the Metaverse may use – motion sensors for gauging user movement, cameras for capturing images, time-of-flight sensors for measuring distances, and many more.

A stream of raw sensor data, however, is very expensive to process [4]. A large portion of this data is ultimately redundant or not useful to the Virtual Service Providers which are constructing the Metaverse. Here, there lies an opportunity to use ML to make the collection of sensor data more efficient.

Semantic information extraction is the process of extracting only the most useful information from raw sensor data. For example, a processor can convert the stream of data from arm movements into a simple "arm movement" signal. This signal can then be sent to the VSP, which can use it to construct a more accurate representation of the user's movement.

In general, this semantic information extraction can be done on the local IoT device. A local ML model would be trained to recognize the critical information.

As the Metaverse continues to become more developed, it is likely that bandwidth requirements will continue to increase. As computer costs fall, the idea of using ML for efficient data compression becomes a more attractive option.

Given the ongoing evolution of the Metaverse and the anticipated increase in bandwidth demands, the utilization of ML for data compression is gaining traction. As the cost of computing decreases, the prospect of employing ML algorithms for efficient data compression becomes increasingly appealing, potentially offering significant benefits in various data transmission applications within the emerging Metaverse landscape.

5.6 DT AND THE METAVERSE

Suppose a VSP wishes to create a digital representation of a physical environment. In order to create an accurate representation, the VSP must stream data from an actual environment to the virtual environment through the use of IoT devices.

The flow of data follows a short path. First, the IoT devices collect data from the physical environment. This data is uploaded to a server that in real time will process the data to create a DT of the environment. The information about the DT is then passed onto several Metaverse users.

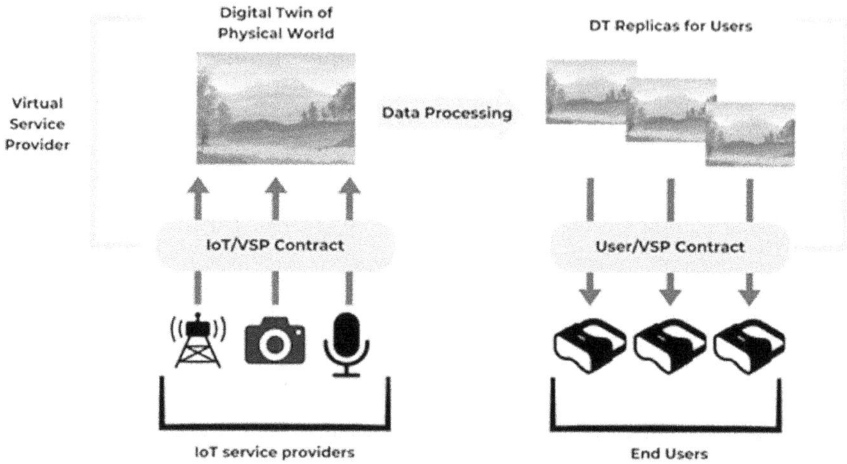

The flow of IoT data in the Metaverse. A VSP server buys data through a contract from IoT devices and constructs a Digital Twin. Data about the twin is sold to Metaverse users.

These DTs will be an essential component of the Metaverse. Users will want to feel as if they are in a real environment, and this means that data from the users and other devices must be collected.

For example, a user may want to virtually shop for a new pair of shoes. The user will want to interact with the store environment, try on the shoes, and see how they appear. The VSP must be able to collect data from the user's movements, the store environment, and the shoes themselves. If such data is collected and processed in real time, the user can have a fully immersive shopping experience in the comfort of their own home. DTs will serve as a cornerstone of many Metaverse applications.

To obtain a highly accurate DT, the VSP must collect a large amount of data from the IoT devices. Managing such a large amount of data is not easy and requires a large amount of bandwidth and computational power.

One approach that the VSP can take to reduce the amount of data required is to use the semantic information extraction principle. This is one of the easiest ways for the VSP to reduce the amount of processing power and network bandwidth it must utilize. By offloading the processing to the IoT devices, the VSP can vastly reduce its expenditure and remain profitable.

5.7 CONTRACT THEORY AND THE METAVERSE

VSPs must operate within their economic constraints. In order to create and manage DTs, the VSP must be able to allocate resources to users in a way that is both efficient and profitable. The VSP must be able to determine the value of IoT devices and users and make transactions accordingly.

A paper by Lofti et al. [5] proposes a method for the VSP to manage the data flow from IoT devices. They identify how to identify the value and utility of end-users and IoT devices and how to allocate resources accordingly. They use a Markov Decision Process (MDP) to determine the best way to allocate resources.

A user who wants to participate in a virtual service would want to be able to measure the quality of any end-product DT. The two aspects that are generally most important are the refresh rate of the DT and the amount of detail it contains. A user who wants a higher refresh rate and a higher quality would be a high-value user because they will buy more services. We can, therefore, define the value of a particular Metaverse user for the VSP as a function of the refresh rate h and resolution r:

$$V(\tau,\phi,r,h) = \tau f_1(r) + \phi f_2(h) \tag{5.2}$$

Here, τ and ϕ are just constants, and $f(r)$ is a function which can weight the value of each individual parameter. The VSP can then use this function to determine the value of a particular user and allocate resources accordingly.

Since the VSP must also provide downstream data for the user, the user also presents a cost in terms of how much data they need. For example, a user who needs higher resolution/higher refresh rate based on their hardware will present a higher cost to the VSP. Therefore, a utility metric can be constructed based on the value metric, where the utility U constitutes both value and cost:

$$U = V - \pi \tag{5.3}$$

Here, π is the cost the user pays. The VSP can then use this utility metric to determine the best way to allocate resources to the user. A higher utility is better of course, but too high a cost would deter users from ever using the service. In order to maximize profitability for the VSP, it must use a more complex algorithm to determine pricing.

In the opposite channel, the VSP must also buy real-time data from IoT devices. In general, a VSP will "hire" IoT devices that collect useful semantic data from the physical world. There are several parameters that characterize the value of any particular IoT device.

An IoT device with a high value will provide a greater set of semantic features. One can quantify the "score" of the different semantic features available as a number. We will use ϕ to represent the semantic score of an IoT device.

In addition to semantic score, the quality of the data transmission link is crucial. Noisy channels will lower the value of the IoT device. A data transmission rate is the second quantity that the VSP considers and is denoted by λ.

Finally, the age of the information is crucial for high-quality user experiences. Devices with a higher refresh rate will provide far greater value to the VSP. We denote the refresh rate as γ.

Complete with this information, the VSP can gauge the total utility of an IoT device used for digital twinning. The utility is the user market value of the data produced $\pi_{\lambda,\gamma,\phi}$ minus the cost of the device's data Y:

$$U = \pi_{\lambda,\gamma,\phi} - Y \tag{5.4}$$

The major problem, however, is that the VSP does not have perfect information about the IoT devices. In particular, ϕ, λ, and γ are all unknown to the VSP, or the VSP can only get a rough gauge on their true values. The challenge presented here is to accurately determine the utility of an IoT device without directly measuring these parameters.

Additionally, the VSP needs to avoid what are called violations – Incentive Compatibility (IC) and Individual Rationality (IR). IC means that the VSP must not be able to manipulate the IoT devices into providing false information. IR means that the IoT devices must not be able to manipulate the VSP into providing a higher reward than they deserve.

The paper proposes, therefore, to predict the best IoT devices to use for digital twinning, through a Markov Decision Process chain. First, the VSP can create a set of contract bundles – each bundle has an associated price, semantic score, data transmission rate, and refresh rate.

The MPD is then able to change either the price or the amount of semantic information it collects – this serves as the "action" from the MPD. The MPD is trained to use information about data transmission, detected IC violations, and detected IR violations to determine how to price a contract bundle that an IoT device accepted.

By doing this, the VSP is able to maximize the utility of the IoT devices it uses for the twinning. A similar process takes place on the VSP/user interface, where users are buying data from the VSP. Each user will have different demands for refresh rate and resolution, and the VSP should allocate resources and price contracts accordingly.

The use of software-based contracts is a key part of the communication within the Metaverse. Engineers should be concerned not only with transmitting data but also with determining the value of data that is being transmitted.

The economics of the Metaverse will be based on data movement. Contract algorithms such as the one posed will be crucial for any VSP to operate efficiently and profitably. By fostering an ideal environment for a data economy to thrive, the Metaverse will be able to grow and expand in ways that are currently unimaginable.

5.8 CELLULAR NETWORKING IN THE METAVERSE

The Metaverse will require a new generation of wireless communication technologies to function. The bandwidth offered by 4G networks is insufficient to handle the intense demand of the Metaverse. High resolution virtual holograms, for example, are estimated to take up 10 times the amount of bandwidth compared to video streaming.

However, we are seeing a rapid evolution in wireless communication. The demand for low-latency and high-bandwidth streaming in mobile devices has opened the door for the kind of technology necessary to support the Metaverse. 5G networking and beyond will be crucial for the Metaverse to function, and many ongoing research projects will ultimately contribute to the success of 5G and future network generations.

5.9 HOW 5G WILL ENABLE THE METAVERSE

Currently, the large majority of cellular devices rely on either 4G or 5G communication standards. 5G networks, by definition, are cellular networks which follow a standard known as the 5th Generation New Radio.

The standards for 5G New Radio are defined by several organizations, including the International Telecommunication Union (ITU), the International Standards Organization (ISO), the Institute of Electrical and Electronics Engineers (IEEE), and, most primarily, the 3rd Generation Partnership Project (3GPP) [6].

6G communication is likely to be the standard following 5G. Its specific characteristics are comparatively less well defined, but many organizations have made proposals for what 6G communication should look like. It is likely that high demand for certain features will influence the development of 6G communication – for example, if Metaverse-specific applications become more popular in the near future, it could encourage the development of 6G communication.

5G is designed to overtake 4G using a variety of technologies. Many more experimental technologies such as Massive Multi-Input Multi-Output) (MIMO) and Beamforming were used in 4G, but now take center stage in newer networks.

Massive MIMO is a communication technology that uses a large array of antennas to send data to multiple users simultaneously. Massive MIMO is based on several technologies, including beamforming, precoding, and spatial multiplexing.

Beamforming is the process of directing a wireless signal at a specific destination. This is achieved through the use of constructive interference. An array of antennas can be used for this purpose – by adjusting the phase of each antenna, the signal can be directed at a specific angle.

Table 5.1 shows the cellular technologies and it's definition and purposes. Massive MIMO is capable of multiplexing, meaning it is capable of transmitting and receiving data from multiple users simultaneously. Multiple beams can be constructed using multiplexing techniques to achieve multi-user communication.

Table 5.1. Cellular technologies

Generation	*Technology*	*Definition*	*Purpose*
4G	Low Band	EM spectrum covering frequency less than 1GHz	Provides broad coverage with good penetration through obstacles
5G	Mid Band	EM spectrum covering frequency between 1GHz and 6GHz	Balances coverage and capacity, suitable for urban and suburban areas
5G	Millimeter Wave	EM spectrum covering frequency above 24GHz	Offers high throughput for dense urban areas and indoor environments
5G	Massive MIMO	Broad term for arrays with dozens of sender/receiver antennas	Enables spatial multiplexing and beamforming for improved spectral efficiency and capacity
5G	Beamforming	Use of constructive interference to direct a wireless signal in a particular direction	Improves signal quality, increases coverage, and enables efficient use of spectrum
5G	SDMA	Space-Division Multiple Access – division of users into spatial domains to share the same frequency band	Increases spectral efficiency and capacity
6G	ELAA	Extremely Large-Scale Antenna Array – Term for arrays with huge amounts of antennas	Enables high data rates and low latency for a large number of users
6G	RHS	Reconfigurable Holographic Surface – thin surface antenna with metamaterials	Enables ELAA capabilities in a cost-effective manner

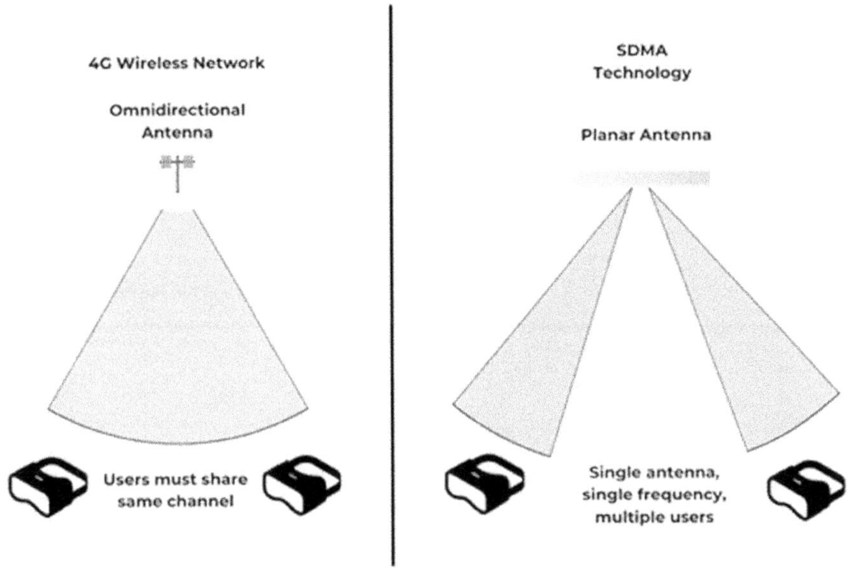

Space Division Multiple Access (SDMA) allows many users to share a frequency band by separating antenna signals in space. This is achieved through the use of beamforming and other techniques.

Beamforming is related to Space-Division Multiple Access (SDMA) techniques for achieving high throughput. In general, any antenna technology that separates users in space and uses the same frequency band is considered SDMA. For the substantial increase in throughput demanded by the Metaverse, novel SDMA technologies will be required. SDMA strategies continue to be an active area of research and will be expanded upon in upcoming 6G standards [7].

The aforementioned technologies will only be expanded upon in 6G and other future communication protocols. One potential technology that 6G can capitalize on is the Extremely Large-Scale Antenna Array (ELAA). Reaching performance metrics superior to 5G requires emitters on the order of hundreds or thousands of elements [8].

5.10 RECONFIGURABLE HOLOGRAPHIC SURFACES

The reconfigurable holographic surface (RHS) is a promising strategy for cheap and effective 6G ELAA antennas. The RHS is a thin surface antenna that is layered with metamaterials. These metamaterials are capable of manipulating electromagnetic waves in ways that traditional antennas cannot. The original authors of the RHS technology pose it as a solution to achieving massive MIMO capabilities in a cost-effective manner [9].

An RHS antenna consists of waveguides, feeds, and metamaterial radiation elements. Each feed is designed to generate a reference wave containing the user signal. The reference propagates in the waveguide, and the metamaterials are used to manipulate the wave into a desired shape.

The radiation elements form a grid on the surface of the RHS. The electromagnetic response of each element is controlled, allowing a beam to be directed at a specific angle.

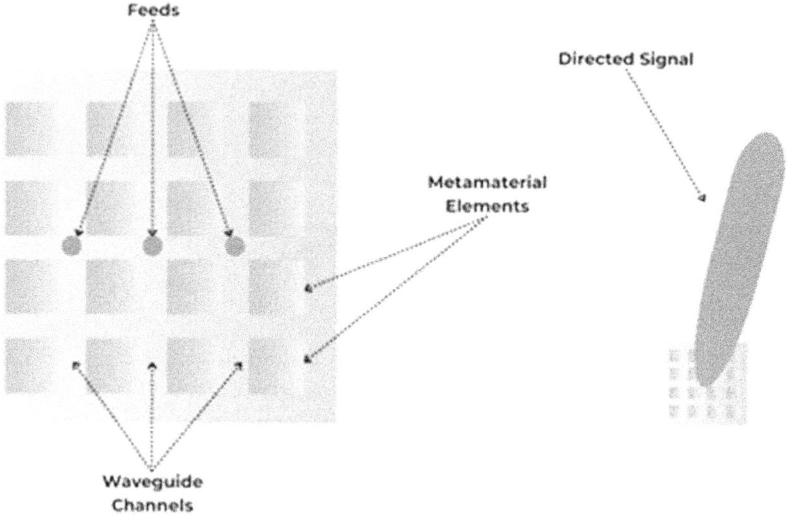

Rough depiction of an RHS antenna, showing the waveguide, feed, and radiation elements.

More specifically, the reference wave is a complex number Ψ_{ref} [10]. The wave generated by feed k at radiation element n_y, n_z is given by Equation (5.5):

$$\Psi_{ref} = e^{-j \cdot \mathrm{k}_s \cdot \mathrm{r}_{n_y, n_z}} \tag{5.5}$$

Here, k_s is known as the propagation vector – a property of the reference wave, and r_{n_y, n_z} is the distance vector of the radiation element.

Then, the object wave Ψ_{obj} is the wave that propagates from the radiation element to the user. This wave is given by Equation (5.6):

$$\Psi_{obj} = e^{-j \cdot \mathrm{k}_f \cdot \mathrm{r}_{n_y, n_z}} \tag{5.6}$$

where k_f is the vector representing the direction of desired propagation.

With this information we use what is known as the holographic inference principle. It tells us that the interference pattern between the reference and object wave is determined by Equation (5.7):

$$\Psi_{int} = \Psi_{obj} \cdot \Psi^{*}_{ref} \tag{5.7}$$

In RHS, if this particular interference pattern is constructed, the radiation elements will then be able to create a beam at a specific angle. Further math would yield exactly how to set each radiation element's response to create the interference pattern. This would, in turn, mean that configuring the RHS to create a beam at a specific angle can be done through a simple calculation.

The RHS is capable of creating multiple beams simultaneously – this means that a single antenna would be capable of reaching a large number of users simultaneously.

The other major advantage of RHS is the size of the antenna. Because the antenna is a thin surface, it would be able to fit into a variety of different form factors. VR headsets and IoT devices, which need to transmit data to more centralized servers, could use a network of RHS antennas, which minimize interference with other devices and open up extra bandwidth in the network.

RHS is a promising beamforming technology that will allow for high-bandwidth communication in the Metaverse. There are many schemes that are currently being developed to build on this and other similar metamaterial technologies for antenna communication.

5.11 SIMULTANEOUS LOCALIZATION AND MAPPING

Antennas are largely thought of as a means of communication. However, the communication technologies that support the Metaverse can also be the same technologies that collect data for it.

Simultaneous localization and mapping (SLAM) is a technology that uses a sensor to map an environment in real time. SLAM is used to create a map of the local environment by directing beams, such as radar beams, in a particular direction. The time it takes for the reflection to return to the sensor is used to determine the distance of the object. With enough information, the object can determine both its relative location and the surrounding environment. The precision and low cost of RHS antennas make them an excellent use case for SLAM.

Zhang [11] describes a related application of SLAM in autonomous vehicle driving. They propose that the use of RHS antennas would be a cheap and reliable way for autonomous vehicles to map their environment.

In their paper, they describe how a network of vehicles, which would be equipped with RHS antennas, would create a more accurate environment map when compared with simple individual vehicle mapping.

Similar technology, if embedded inside IoT devices or end-user products, could directly be applied in the Metaverse. With compact RHS antennas, the end user can generate useful data about their environment and craft a more immersive experience for themselves and other users.

5.12 CONCLUSION

The Metaverse is promising to become a new medium for human interaction. Such a large network of devices will be built on a new generation of communication technologies.

To fully realize the Metaverse, the communication hardware and software that underpins it must be improved. The Metaverse will require high-bandwidth, low-latency communication technologies to function. Many researchers are working to develop new protocols, hardware, and software that will be directly applicable to the Metaverse.

In the software realm, the recent advances of ML have shown great promise in revolutionizing software advances. In the field of communication, ML algorithms can offer novel ways to extract useful information from large amounts of data and reduce the amount of data required for transmission.

The management of data flow between users, devices, and servers is a crucial part of the Metaverse. The use of ML algorithms combined with many innovative software engineering techniques can help to optimize the data flow and make communication in the Metaverse more efficient.

In the hardware realm, the development of new communication technologies is crucial for the Metaverse to function. Massive MIMO technologies, such as the Reconfigurable Holographic Surface, are promising to enable high-bandwidth communication in the Metaverse. These technologies will bring about the next generation of wireless communication and enable a large number of small devices to share network space in an effective manner.

With further research and development, a diverse array of research teams in different paradigms will help enable the wireless technology and communication infrastructure for virtual reality to fully thrive. The Metaverse will be built on the communication infrastructure that is being discovered and developed today.

REFERENCES

1. Facebook Business. (2021). The Metaverse: A New Frontier for Business. *Facebook Business*. https://www.facebook.com/business/news/insights/metaverse-a-new-frontier-for-business.
2. Bruer, Alexandra. (2021, November 8). *Mapping the International 5G Standards Landscape and How It Impacts U.S. Strategy and Policy.* Informative Technology & Innovation Foundation (ITIF).
3. Fujihashi, T. (2023). Soft 2D-to-3D Delivery Using Deep Graph Neural Networks for Holographic-Type Communication. Proceedings of the IEEE International Conference on Acoustics, Speech, and Signal Processing (ICASSP).
4. Westphal, C. (2023, December). Networking for the Metaverse: The Standardization Landscape. arXiv:2312.09295v1 [cs.NI]. https://arxiv.org/abs/2312.09295v1
5. Hamidouche, W., et al. (2023). Immersive Media and Massive Twinning: Advancing Towards the Metaverse. *arXiv e-print*, arXiv:2307.01522.
6. Lotfi, I. (2023). Semantic Information Marketing in the Metaverse: A Learning-Based Contract Theory Framework. *arXiv e-print*, arXiv:2302.11457.
7. Zidong Wu and Linglong Dai. (2023, June 27). *Multiple Access for Near-Field Communications: SDMA or LDMA?*. arXiv preprint arXiv:2208.06349. https://arxiv.org/pdf/2208.06349.
8. Mingyao Cui et al. (2022). *Near-Field Communications for 6G: Fundamentals, Challenges, Potentials, and Future Directions*. arXiv preprint arXiv:2203.16318. https://arxiv.org/pdf/2203.16318.
9. Haobo Zhang. (2023, March 24). *Holographic Integrated Sensing and Communications: Principles, Technology, and Implementation*. arXiv preprint arXiv:2211.08646. https://arxiv.org/pdf/2211.08646.
10. Deng, R., et al. (2021) Reconfigurable Holographic Surfaces for Future Wireless Communications. *arXiv e-print*, arXiv:2112.06372.
11. Zhang, H. (2023). *Reconfigurable Holographic Surface Aided Collaborative Wireless SLAM Using Federated Learning for Autonomous Driving*. IEEE.

Chapter 6

Virtual Hearts

Navigating Love, Faith, and Ethics in the Age of 3D Virtual Humans in the Metaverse

Timothy Adewale Adeyi, Adrian David Cheok, Jun Cai, and Ying Yan

6.1 INTRODUCTION

The advent of the Metaverse and the proliferation of 3D avatars mark a significant paradigm shift in the digital landscape, transcending traditional boundaries and ushering in a new era of interconnected virtual experiences. The concept "Metaverse" clearly states that the convergence of physical and virtual reality (VR) creates the collective virtual shared space. This dynamic digital environment is inhabited by users in the form of 3D avatars, representing a synthesis of real-world interaction and digital immersion. Exploring the intricate tapestry of the Metaverse, becomes imperative in explaining the origins and evolution of this conceptualization.

The concept of the Metaverse finds its roots in science fiction, with Neal Stephenson's seminal novel *Snow Crash* [1] presenting a vivid portrayal of a VR space where users interact through avatars. Over the years, advancements in technology, particularly in VR and augmented reality (AR), have brought the Metaverse from the realms of fiction to tangible digital existence. Companies such as Meta (formerly Facebook), Epic Games, and Decentraland are at the forefront of shaping this immersive digital frontier, providing platforms where users engage with each other in a three-dimensional space.

Central to the Metaverse experience are 3D virtual human avatars, digital representations of users that serve as their virtual embodiments within the digital realm. These avatars are customizable, allowing individuals express their identity and personality in a way that extends beyond the limitations of the physical world. The rise of 3D avatars is intricately linked to the technological advancements in motion capture, computer graphics, and artificial intelligence. This convergence of technologies enables the creation of highly realistic and responsive avatars which enhances the sense of presence and social interaction in the Metaverse.

Considering the influence of 3D avatars on the Metaverse, it is essential to recognize their role to allow social connections and facilitate meaningful interactions. The ability to express emotions, engage in non-verbal communication, and navigate virtual spaces through avatars adds a layer of depth

 DOI: 10.1201/9781032663609-7

to online interactions that was previously unattainable. Scholars such as J. N. Bailenson, J. Blascovich, and R. E. Guadagno have explored the concept of "digital co-presence," emphasizing how the use of avatars can create a sense of being physically present with others in a shared virtual space.

However, the rise of 3D avatars in the Metaverse has profound implications for industries beyond gaming and social networking. The business landscape is witnessing a transformation, with virtual conferences, trade shows, and collaborative workspaces becoming commonplace. This paradigm shift is undermined by the work of Reeves and Nass [3], who demonstrated the impact of social cues in computer-mediated communication, laying the foundation for understanding the significance of avatars in virtual environments.

However, contemplating the multifaceted dimensions of the Metaverse and the ascent of 3D avatars is cleared enough as a cusp of a digital renaissance. The continuous integration of virtual and physical realities, coupled with the expressive potential of 3D avatars, predicts a future where the distinctions between reality and the virtual realm become indistinct. Comprehension of the full scope of the transformative phenomenon is imperative in exploring the intricacies of virtual embodiment, technological innovation, and the societal implications that converge in the dynamic landscape of the Metaverse.

6.2 STATEMENT OF PROBLEM

In the digital age, a perplexing phenomenon has emerged – romantic affection directed toward non-human entities within digital spaces. As individuals increasingly form emotional bonds with virtual characters, AI companions, or even inanimate objects, questions arise about the nature and implications of such attachments. This shift prompts an exploration into the psychological and sociocultural factors driving these affections, considering the role of technology in shaping human emotions. Scholars like Turkle [4], have explore the complexities of human-technology interactions, explaining the evolving nature of relationships in a digitally mediated world. Understanding this novel aspect is crucial for navigating the intricate intersections of technology and human emotion.

6.3 OBJECTIVE OF THE STUDY

The main objective of this research work is to critically examine the moral and religious dimensions of human emotional attachment to virtual entities. We investigate how such affections align with or challenge moral frameworks and religious beliefs providing a more valuable insights into the evolving intersection of technology and spirituality. This research study aims in contributing a more subtle understanding of the ethical implications and theological considerations surrounding the interaction with 3D avatars, enriching the discourse on the evolving dynamics between digital culture and traditional belief systems.

6.4 LITERATURE REVIEW

6.4.1 Historical Understanding of Human Relationships with Artificial Entities

The historical understanding of human relationships with artificial entities, particularly in this contextual study of affection toward 3D avatars, explores the evolving nature of our connections with non-human entities. Examining this history unveils the intricate tapestry of human interaction with artificial entities, shedding light on the moral, religious, and biblical dimensions that have shaped our perceptions.

Historically, the roots of human interaction with artificial entities can be traced back to ancient mythology and folklore, where tales of animated statues, golems, and enchanted objects abound. These narratives often portrayed humanity's fascination and, at times, trepidation toward creations brought to life. In exploring these historical narratives, we find early inklings of the complex relationship humans have had with the artificial, suggesting an enduring fascination with imbuing life into non-living forms [5, 6].

As societies progressed, the Renaissance era witnessed a resurgence of interest in automata and artificial life. Mechanical marvels, such as clockwork automata, captivated the imaginations of scholars and artists. This era marked a shift in the perception of artificial entities from mere curiosities to reflections of human ingenuity. The works of inventors like Leonardo da Vinci and Jacques de Vaucanson exemplify this era's fascination with creating lifelike automata, laying the groundwork for future explorations into the emotional connection between humans and artificial beings [7].

The Industrial Revolution brought forth further advancements in automaton technology, with the development of more sophisticated and realistic machines. However, this period also concerns the emergence of the potentiality in dehumanizing effects of technology on society. The anxieties expressed by thinkers such as Mary Shelley in *Frankenstein* [8] reflected broader societal worries about the moral consequences of creating artificial life and the responsibilities that come with it [8].

Moving into the 20th century, the rise of science-fiction literature and cinema explored the theme of human relationships with artificial beings in more nuanced ways. Works such as Isaac Asimov's *I, Robot* (1950) introduced the concept of robots governed by ethical rules, raising questions about the moral implications of creating intelligent machines. As technology advanced, these fictional explorations foreshadowed real-world dilemmas and debates surrounding the ethical treatment of artificial entities [9].

In the contemporary era, the introduction of the digital age and the rise of 3D virtual human avatars have ushered in a new chapter in the historical narrative of human interaction with artificial entities. Virtual environments, social media, and online gaming platforms provide spaces where individuals form emotional attachments to digital representations of themselves and others. This shift prompts a reassessment of historical perspectives, with scholars such as M. Arnd-Caddigan [4] examining the implications of these digital relationships on human identity and connection in the 21st century.

In conclusion, the historical understanding of human relationships with artificial entities reveals a nuanced journey marked by fascination, caution, and ethical considerations. From ancient myths to modern digital realms, the intricate interplay between humans and artificial beings continues to shape our moral, religious, and biblical perspectives on affection toward 3D avatars.

6.4.2 Previous Works on Human-Computer Interactions and Feelings

Previous works on human-computer interactions and feelings of affection toward 3D humans have explored the intricate dynamics of emotional connections in the digital realm. The field of human-computer interaction (HCI) has witnessed a rich body of investigation on how individuals perceive and relate to artificial entities, shedding light on the psychological and emotional aspects of these interactions.

Scholars in the realm of HCI have extensively examined the concept of presence, emphasizing the degree to which users feel immersed in virtual environments [10]. This research has laid the foundation for understanding the emotional impact of 3D avatars, as their realism contributes to a heightened sense of presence, influencing the depth of emotional engagement. Moreover, studies within HCI have delved into the phenomenon of parasocial interaction, where individuals form emotional connections with media characters, including those represented by 3D avatars [11, 12]. This framework provides insights into the nature of affectionate feelings toward virtual entities, offering a lens through which to understand the emotional bonds users may develop with 3D avatars.

The exploration of anthropomorphism in HCI literature is also pertinent to the study. The tendency to ascribe human characteristics to entities that are not human, such as 3D avatars, has been widely examined [3] . Understanding how individuals anthropomorphize these digital entities contributes to unraveling the complexities of emotional connections, as users may project human emotions onto their avatars.

Furthermore, research on the intersection of technology and psychology has probed the impact of social cues on human behavior in digital spaces [2]. The incorporation of social cues through 3D avatars enhances the emotional richness of online interactions, influencing users' feelings of connection and affection toward their virtual representations.

In sum, previous works in HCI and related fields offer valuable insights into the intricate tapestry of human emotions in digital spaces, especially concerning affection toward 3D avatars. These studies provide a foundation for exploring the moral, religious, and biblical dimensions of these emotional connections in the context of the current research.

6.4.3 Religious Perspective on Non-human Affection: From Statues and Symbols to Digital Entities

Examining as an example religion, Christian definitions of love and marriage in the context of affection toward 3D avatars involves a nuanced exploration of biblical perspectives. Christian teachings often emphasize the centrality of love in human relationships, drawing inspiration from passages like 1 Corinthians 13, known as the "Love Chapter." This biblical definition of love encompasses aspects such as patience, kindness, and selflessness, which are traditionally associated with human-to-human connections [13].

In the context of 3D avatars, the application of Christian definitions of love prompts contemplation on the nature of affectionate bonds formed in virtual spaces. While the Bible provides guidance on human relationships, the ethical considerations surrounding the extension of these principles to digital entities remain complex. Christians may grapple with questions about the capacity of avatars to reciprocate love in the same manner as humans and whether these digital connections align with the biblical understanding of love as a sacrificial and mutual endeavor.

Moreover, the examination of Christian definitions of marriage is pertinent, as marriage is often regarded as a sacred union in Christianity. The biblical concept of marriage, rooted in passages such as Genesis 2:24, emphasizes the unity and companionship between a man and a woman. Applying these teachings to affection toward 3D avatars raises questions about the alignment of virtual relationships with the sacredness and purpose ascribed to traditional marital bonds [14].

Christian perspectives on marriage often highlight procreation and the family unit as integral components of the divine plan. The ethical implications of extending affection to avatars, which do not partake in the biological aspects of human relationships, may be scrutinized within the Christian framework. The examination of Christian definitions of love and marriage provides a foundation for evaluating the moral dimensions of affection toward 3D avatars from a religious perspective, facilitating a deeper exploration of the ethical considerations that arise within the intersection of technology and faith.

6.4.4 The Bible Passages That Indirectly Relate to Human-avatar Relationships

Exploring Bible passages that indirectly relate to human-avatar relationships requires a careful examination of biblical principles that may offer insights into the moral dimensions of affection toward 3D avatars. While the Bible does not explicitly address virtual relationships, certain passages provide principles that can be applied to contemporary digital interactions.

One relevant aspect is the biblical emphasis on idolatry. Throughout the Old Testament, the Bible warns against the worship of false gods and the creation of graven images (Exodus 20:3-4). While avatars may not be worshiped, the parallels between creating digital representations and the caution against idolatry invite reflection on the ethical boundaries of affection toward avatars. The biblical stance against idolatry suggests a cautionary approach to the emotional investment in digital entities, prompting believers to consider whether the affection directed toward avatars aligns with the biblical prohibition of idolatrous practices.

Additionally, the biblical teaching on the sanctity of human life may indirectly intersect with the virtual realm. The Bible underscores the inherent value of each human being as they are created in the image of God (Genesis 1:27). This principle may prompt considerations about the ethical treatment of avatars, questioning the extent to which virtual entities should be afforded moral consideration. The biblical perspective on the sanctity of human life challenges believers to critically evaluate their emotional connections with avatars, considering whether these interactions uphold the inherent dignity of humanity.

Moreover, passages such as Mark 12:31 that emphasize the commandment to love one's neighbor as oneself thus provide a moral framework that extends to digital interactions. Applying this principle to human-avatar relationships encourages believers to approach virtual connections with the same love, respect, and ethical considerations that guide their interactions in the physical world.

Examining these indirect connections between biblical passages and human-avatar relationships provides a comprehensive foundation for moral reflections within a religious context. While the Bible may not explicitly address digital interactions, its overarching principles can guide believers in navigating the evolving landscape of technology and relationships.

6.5 THE THEOLOGICAL INTERPRETATIONS AND DEBATES REGARDING THE MORALITY OF HUMAN-AVATAR UNIONS

Theological interpretations and debates surrounding the morality of human-avatar unions delve into the intricate intersections of technology, spirituality, and ethics. In theological discourse, the question arises whether forming emotional bonds with 3D avatars is morally acceptable or whether it challenges traditional views on relationships and human purpose.

From a theological standpoint, the concept of imago Dei, the belief that humans are created in the image of God, raises questions about the extension of emotional connections to digital entities. Theological debates may revolve around whether avatars, as products of human creation, possess any inherent spiritual significance or if the emphasis should remain on human-to-human relationships as reflections of the divine image (Genesis 1:27).

Furthermore, discussions on the morality of human-avatar unions may intersect with theological considerations regarding the purpose of human relationships. Theologians may reflect on biblical teachings that emphasize companionship, such as the creation of Eve as a suitable helper for Adam (Genesis 2:18). Debates may arise as to whether the divine intention for human relationships aligns with or challenges the emotional bonds formed with virtual entities.

The evolving nature of technology prompts inquiries into theological views on the soul and consciousness. Theological interpretations may grapple with whether avatars possess any form of consciousness or if the emotional connections formed with them have spiritual implications. This consideration delves into profound theological debates on the nature of existence and the boundaries of spiritual significance.

While theological perspectives on the morality of human-avatar unions may differ among scholars and religious traditions, the discussions provide an avenue for reflection on the evolving dynamics between technology and faith. Theological discussions enhance our comprehension of moral implications of affection toward 3D avatars and its alignment with or deviation from established religious principles.

6.6 DETAILED DESCRIPTION OF QUALITATIVE METHODS USED

The exploration of affection toward 3D avatars from the lenses of morality, religion, and biblical perspectives requires a nuanced qualitative research approach. In this study, qualitative methods play a pivotal role in capturing the depth and complexity of individuals' experiences and beliefs.

Qualitative interviews constitute a primary method, allowing researchers to engage participants in open-ended conversations. These interviews aim to elicit rich narratives about participants' emotional connections with 3D avatars, uncovering the moral considerations and religious nuances that shape their perspectives [15, 16]. The open-ended nature of qualitative interviews facilitates a detailed exploration of participants' thoughts, feelings, and the intricacies of their beliefs regarding the moral dimensions of virtual relationships.

Additionally, thematic analysis serves as a robust qualitative method to identify recurrent themes within participants' responses [17]. This approach enables researchers to uncover patterns related to moral reasoning, religious influences, and biblical interpretations. Thematic analysis contributes to the comprehensive understanding of how individuals conceptualize affection toward 3D avatars within the context of their moral and religious frameworks.

The qualitative study also involves participant observation within virtual environments where individuals interact with avatars. This method allows researchers to observe firsthand the dynamics of these digital relationships, gaining insights into the emotional investment, symbolic actions, and interactions that unfold within the virtual space [18]. Participant observation enhances the depth of the study by providing a contextual understanding of how affection toward avatars manifests in participants' behaviors.

Furthermore, reflexive journaling plays a crucial role in qualitative research on sensitive topics like morality and religion. Researchers maintain reflective journals to document their own biases, preconceptions, and reflections throughout the research process [19]. This reflexivity contributes to the transparency of the study and ensures that the researchers' perspectives do not unduly influence the interpretation of participants' experiences.

In conclusion, the detailed description of qualitative methods in this study involves in-depth interviews, thematic analysis, participant observation, and reflexive journaling. These methods collectively contribute to a comprehensive exploration of affection toward 3D avatars, providing valuable insights into the moral, religious, and biblical dimensions of virtual relationships.

6.7 X EXPLANATION OF TEXTUAL ANALYSIS AND INTERPRETATION STRATEGIES

In the investigation of affection toward 3D human avatars, textual analysis and interpretation strategies offer a robust framework for uncovering nuanced meanings within written or spoken discourse. Textual analysis involves a systematic examination of language, identifying patterns and themes within the data [20]. This method is particularly relevant in exploring the moral and religious dimensions of affection toward avatars, as it allows researchers to dissect participants' narratives and discern the underlying meanings embedded in their language.

Interpretation strategies within textual analysis encompass various approaches to make sense of the data. One common strategy is hermeneutics, which involves a detailed examination of the text to uncover hidden meanings and understand the context in which statements are made [21]. In the context of this research, hermeneutics can aid in deciphering participants' perspectives on the moral and religious aspects of their affection toward 3D avatars, shedding light on the underlying beliefs that inform their expressions.

Moreover, critical discourse analysis (CDA) serves as a valuable strategy within textual analysis. CDA allows researchers to explore power dynamics, ideologies, and societal structures embedded in language [22]. This approach is pertinent when investigating how moral and religious values may be influenced or challenged by digital interactions with avatars. By applying CDA, researchers can unveil the discursive mechanisms through which participants negotiate their moral and religious beliefs within the context of virtual relationships.

To ensure rigor and reliability, triangulation of interpretation strategies may be employed. By using multiple strategies, researchers can cross-verify findings and enhance the overall validity of the interpretations [23]. This methodological approach strengthens the robustness of the study by offering a more detail understanding of the complex interplay between morality, religion, and biblical perspectives in the framework of affection toward 3D avatars.

In conclusion, textual analysis and interpretation strategies provide a methodologically sound approach for investigating the multifaceted dimensions of affection toward 3D avatars. By systematically examining language and employing interpretative strategies, researchers can unlock the intricate meanings embedded in participants' narratives, contributing to a nuanced understanding of the moral and religious implications of virtual relationships.

6.8 ETHICAL CONSIDERATIONS DURING THE RESEARCH

Ethical considerations play a paramount role in research, particularly when delving into sensitive topics such as morality, religion, and biblical perspectives on affection toward 3D avatars. Ensuring the ethical treatment of participants and maintaining the integrity of the research process are fundamental principles guiding this study.

In qualitative research, up-to-date consent is a crucial ethical consideration [24]. Participants should be fully conscious of the study's purpose, potential risks, and benefits before providing their consent to participate. This transparency is essential in studies exploring personal beliefs and emotional experiences, as it empowers participants to make informed decisions about their involvement.

Confidentiality and anonymity further underscore ethical considerations. Researchers must safeguard participants' identities and ensure that their responses cannot be traced back to them. This is especially critical when discussing personal beliefs and moral perspectives, as participants may share sensitive information about their religious views and emotional connections with avatars [25].

Moreover, the potential psychological impact on participants must be carefully considered. Engaging in discussions about deeply held beliefs and emotional connections can evoke strong reactions. Researchers should be vigilant in monitoring participants' well-being throughout the study and provide appropriate support or referrals if needed [24].

Maintaining cultural sensitivity is paramount in studies that involve religious perspectives. Researchers must approach participants with respect for their beliefs, recognizing the diversity of religious views and avoiding any form of bias or imposition of the researcher's views [23].

Triangulation of data sources and interpretations also contributes to ethical rigor. By corroborating findings through multiple methods, researchers can enhance the credibility of the study and ensure that the interpretations accurately represent participants' perspectives [20].

In conclusion, ethical considerations in this research are integral to protecting the rights and well-being of participants, ensuring their autonomy and privacy, and fostering a respectful exploration of their beliefs and experiences. Adhering to ethical principles is fundamental in producing valid and reliable insights into the morality, religion, and biblical perspectives surrounding affection toward3D avatars.

6.9 RESULTS OF TEXTUAL ANALYSIS ON RELEVANT BIBLICAL PASSAGES

In exploring the intricate relationship between affection toward 3D avatars and its intersection with morality, religion, and biblical perspectives, a textual analysis of relevant biblical passages yields insightful findings. One paramount aspect to consider is the biblical concept of creation, as it shapes the foundation for understanding human relationships with artificial entities. Genesis 1:26-27, which articulates that humans beings are created in the image of God, raises questions about the ethical implications of forming emotional connections with non-human entities.

Furthermore, the biblical perspective on idolatry becomes a crucial focal point. Exodus 20:4 admonishes against the worship of graven images, prompting a nuanced examination of whether affection toward 3D avatars may be construed as a form of modern idolatry. This scriptural warning introduces a moral dimension to the discourse, emphasizing the need for ethical considerations in the interaction with virtual entities.

Delving into the New Testament, the teachings of Jesus regarding love and compassion offer a lens through which to evaluate the morality of affection toward 3D avatars. The instruction to love one's neighbor as oneself (Matthew 22:39) prompts reflection on the extension of this love to virtual entities and the moral responsibilities associated with such affections.

Moreover, the biblical story of the Good Samaritan (Luke 10:25-37) provides a paradigm for compassion that transcends traditional boundaries. Applying this narrative to the context of affection toward 3D avatars invites contemplation on the moral imperatives of extending empathy and care beyond the human sphere.

While biblical passages may not explicitly address 3D avatars, a comprehensive analysis allows for the extrapolation of moral and ethical principles. Considering the evolving nature of technology, the interpretation of biblical teachings in contemporary contexts becomes imperative. In this regard, theological scholars such as [26]advocate for a dynamic hermeneutical approach that acknowledges the applicability of biblical principles to emerging ethical dilemmas.

In conclusion, the textual analysis of relevant biblical passages unveils multifaceted dimensions in the discourse on affection toward 3D avatars. The exploration of creation, idolatry, and moral teachings provides a robust foundation for understanding the complexities of human interactions with artificial entities. Engaging with biblical perspectives offers valuable insights into the ethical considerations surrounding affection toward 3D avatars, contributing to a nuanced and informed discussion on the intersection of technology, morality, and religion.

6.10 INTERPRETATIONS AND IMPLICATIONS FOR CHRISTIAN BELIEVERS

Examining the interpretations and implications for Christian believers regarding affection toward 3D avatars unveils a complex interplay of moral, theological, and existential considerations. Within the Christian framework, the interpretation of biblical principles regarding the sanctity of human relationships provides a foundational standpoint (Genesis 2:18-24). This scriptural emphasis on the significance of human connection prompts a reflection on the ethical boundaries of affection directed toward virtual entities.

Moreover, the concept of idolatry, as cautioned against in the Bible (Exodus 20:4), raises pertinent questions for Christian believers. The implications of forming emotional bonds with 3D avatars may be perceived through the lens of potential idolatrous tendencies, necessitating a conscientious evaluation of the nature of such attachments.

Theological perspectives on love and compassion, drawn from the teachings of Jesus, offer a guiding framework for Christian believers navigating the terrain of affection toward 3D avatars. The commandment to love one's neighbor (Matthew 22:39) becomes particularly relevant in contemplating the extension of compassion to virtual entities.

Furthermore, the implications of human agency and ethical responsibility in the creation and interaction with 3D avatars call for a nuanced understanding of stewardship. Christian believers may find resonance with the biblical concept of stewardship, where humanity is entrusted with the care of God's creation (Genesis 2:15), extending this responsibility to the ethical use of technological innovations.

In grappling with these interpretations and implications, Christian scholars advocate for a balanced approach that integrates technological advancements with biblical principles. This underscores the importance of informed, ethical decision-making in the rapidly evolving landscape of human-technology interactions.

In summary, the interpretations and implications for Christian believers regarding affection toward 3D avatars involve a thoughtful engagement with biblical principles on human relationships, idolatry, love, and stewardship. Navigating this terrain requires a discerning approach that aligns technological advancements with the ethical imperatives of the Christian faith.

6.11 BROADER SOCIETAL IMPACTS AND CONSIDERATIONS

The phenomenon of affection toward 3D avatars holds significant broader societal impacts and considerations, particularly when viewed through the lenses of morality, religion, and Biblical perspectives. In exploring these dimensions, it is imperative to acknowledge the evolving nature of human relationships with virtual entities and the ethical implications that ensue. Morality, as a guiding principle, becomes a pivotal aspect in dissecting the implications of affection toward 3D avatars. The moral fabric of society is intricately woven with beliefs and values that shape interpersonal relationships, and the advent of affection directed at virtual entities introduces a novel dimension to this ethical tapestry.

Religion, being a cornerstone of societal norms and values, contributes substantially to the discourse surrounding the affectionate inclinations toward 3D avatars. Various religious doctrines proffer unique perspectives on the sanctity of human connections and the moral boundaries that govern such relationships. As individuals navigate the intricate interplay between their faith and the emerging technological landscape, questions arise about the compatibility of affection for virtual beings with established religious tenets.

Delving into Biblical perspectives unveils a rich tapestry of narratives and moral teachings that form the bedrock of Western ethical frameworks. Examining the scriptures through the prism of affection toward 3D avatars necessitates a nuanced understanding of the biblical stance on love, companionship, and the sanctity of human connections. It is within this scriptural context that the implications of virtual relationships find resonance, prompting a reevaluation of traditional norms in light of contemporary technological advancements.

Furthermore, the societal impacts extend beyond individual beliefs and encompass communal dynamics. The emergence of affectionate bonds with 3D avatars raises questions about the collective moral compass and societal acceptance. As communities grapple with the evolving landscape of human-avatar relationships, discussions on acceptable norms and shared values become imperative for fostering a cohesive societal fabric that accommodates technological advancements without compromising foundational ethical principles.

In conclusion, the broader societal impacts and considerations surrounding affection toward 3D avatars are intricate and multifaceted, encompassing dimensions of morality, religion, and Biblical perspectives. The ethical implications of these evolving relationships necessitate a careful examination of individual and communal values, challenging existing norms and prompting meaningful dialogues on the intersection of technology, morality, and faith within the fabric of society.

6.12 CONCLUSION

This chapter delves into the intricate interplay of human emotions and digital entities within the burgeoning Metaverse. The introduction provides a contextual backdrop, elucidating the rise of 3D avatars and articulating the problem statement: the emergence of romantic affection toward non-human entities in digital spaces. The study's significance is underscored, emphasizing the need to comprehend the implications of evolving human-avatar relationships. The literature review traces the historical trajectory of human interactions with artificial entities, juxtaposing it with contemporary works on human-computer interactions. Notably, the religious dimension is explored, tracing the evolution from traditional symbols to digital entities.

Examining the spectrum of human-avatar relations, the chapter categorizes and parallels emotional connections, exploring the concept of "authentic" affection in virtual spaces. The section on morality and ethics probes philosophical standpoints, distinguishing admiration, affection, love, and obsession in digital realms, and unravels the societal implications and potential issues arising from these relationships. The Christian perspective scrutinizes definitions of love and marriage, analyzes relevant Bible passages, and engages in theological debates about the morality of human-avatar unions.

The methods section elucidates the qualitative research approach, detailing textual analysis and interpretation strategies, while emphasizing ethical considerations throughout the study. Findings and discussions unveil the results of the textual analysis on pertinent biblical passages, offering interpretations and implications for Christian believers. The broader societal impacts and considerations are expounded upon, contextualizing the findings within the larger societal fabric.

In the concluding section, the chapter succinctly summarizes key findings, weaving together insights from the literature review, empirical findings, and theological considerations. The comprehensive exploration of affection toward 3D avatars within the realms of morality, religion, and Biblical perspectives not only contributes to the academic discourse but also prompts contemplation on the evolving nature of human relationships in the digital age.

6.13 THE POTENTIAL FUTURE OF HUMAN-AVATAR RELATIONSHIPS IN THE CHRISTIAN COMMUNITY AND SOCIETY AT LARGE

The potential future of human-avatar relationships within the Christian populace and society at large, as explored in this chapter presents a fascinating intersection of technological advancements and deeply rooted religious beliefs. Within the Christian community, the evolving nature of these relationships prompts contemplation on the compatibility of such unions with traditional definitions of love and marriage. As theological interpretations and debates continue, the question of the moral acceptability of human-avatar connections becomes a central concern for believers navigating the complexities of faith in the digital age.

This future trajectory extends beyond individual beliefs, infiltrating societal norms and structures. The Christian community, as an integral part of society, becomes a focal point for examining the broader implications. The potential societal shifts raise critical questions about the collective moral compass, challenging established norms and prompting discussions on the ethical boundaries of human-avatar relationships. As these connections become more prevalent, society at large must grapple with the integration of technological progress while preserving the cultural and ethical values that underpin social cohesion.

In summary, the potential future of human-avatar relationships within the Christian community and society at large unfolds as a delicate interplay between technological advancements and deeply ingrained religious and societal norms. The ongoing dialogue and reflection within the Christian community contribute to shaping the broader societal perspective on the moral and ethical dimensions of these relationships in the ever-evolving digital landscape.

6.14 RECOMMENDATIONS FOR FURTHER RESEARCH

First, a comprehensive analysis of diverse religious perspectives beyond Christianity would provide a more inclusive understanding of how various faiths interpret human-avatar relationships. Additionally, investigating into the perspectives of religious leaders and scholars could provide subtle insights into the changing moral terrain. Furthermore, studying the influence of cultural variations on the view of these relationships would contribute to a more comprehensive cross-cultural analysis. Such avenues would enrich the discourse, providing a more holistic understanding of the multifaceted interplay between technology, morality, and religious beliefs.

REFERENCES

1. N. Stephenson, *Snow Crash*. Bantam Books, 1992.
2. J. N. Bailenson, J. Blascovich, and R. E. Guadagno, "Self-representations in immersive virtual environments 1," *Journal of Applied Social Psychology*, vol. 38, no. 11, pp. 2673–2690, 2008.

3. B. Reeves and C. Nass, *The Media Equation: How People Treat Computers, Television, and New Media Like Real People*, vol. 10, no. 10. Cambridge University Press, 1996.
4. M. Arnd-Caddigan, *Sherry Turkle: Alone Together: Why We Expect More from Technology and Less from Each Other.* Basic Books, 2011, 348 pp, ISBN 978-0465031467 (pbk), 2015.
5. J. Campbell, *The Hero with a Thousand Faces*, vol. 17. New World Library, 2008.
6. C. Timmermann, H. Kettner, C. Letheby, L. Roseman, F. E. Rosas, and R. L. Carhart-Harris, "Psychedelics alter metaphysical beliefs," *Scientific Reports*, vol. 11, no. 1, p. 22166, 2021.
7. A. Picon, *Digital Culture in Architecture*, Birkhäuser, 2010.
8. M. Shelley, *Frankenstein.* Lackington, Hughes, Harding, Mavor, Jones, 1818.
9. L. Meinecke and L. Voss, "I robot, you unemployed: robotics in science fiction and media discourse," *Schafft Wissen. Gemeinsames und geteiltes Wissen in Wissenschaft und Technik*, vol. 2, pp. 203–221, 2018.
10. N. C. Nilsson, S. Serafin, F. Steinicke, and R. Nordahl, "Natural walking in virtual reality: A review," *Computers in Entertainment (CIE)*, vol. 16, no. 2, pp. 1–22, 2018.
11. D. Horton and R. Richard Wohl, "Mass communication and para-social interaction: Observations on intimacy at a distance," *Psychiatry*, vol. 19, no. 3, pp. 215–229, 1956.
12. W. M. Lim, S. Kumar, S. Verma, and R. Chaturvedi, "Alexa, what do we know about conversational commerce? Insights from a systematic literature review," *Psychology & Marketing*, vol. 39, no. 6, pp. 1129–1155, 2022.
13. I. J. Martin III, "I corinthians 13 interpreted by its context," *Journal of Bible and Religion*, vol. XVIII, no. 2, pp. 101–105, 1950.
14. M. Warner, "'Therefore a man leaves his father and his mother and clings to his wife": Marriage and intermarriage in genesis 2: 24," *Journal of Biblical Literature*, vol. 136, no. 2, pp. 269–288, 2017.
15. J. A. Smith, P. Flowers, and M. Larkin, "Interpretative phenomenological analysis: Theory, method and research," *Qualitative Research in Psychology*, vol. 6, no. 4, 2009.
16. I. Tuffour, "A critical overview of interpretative phenomenological analysis: A contemporary qualitative research approach," *Journal of Healthcare Communications*, vol. 2, no. 4, p. 52, 2017.
17. V. Braun and V. Clarke, "Using thematic analysis in psychology qualitative research in psychology," *Qualitative Research in Psychology*, vol. 3, no. 2, pp. 77–101, 2006.
18. P. Atkinson, *Ethnography: Principles in Practice.* Routledge, 2007.
19. L. Finlay, "Negotiating the swamp: The opportunity and challenge of reflexivity in research practice," *Qualitative Research*, vol. 2, no. 2, pp. 209–230, 2002.
20. J. W. Creswell and J. D. Creswell, *Research Design: Qualitative, Quantitative, and Mixed Methods Approaches.* Sage Publications, 2017.
21. G. Gadamer, J. Weinsheimer, and D. Marshall, *Truth and Method.* Bloomsbury Academic, 2013.
22. N. Fairclough, *Language and Power.* Routledge, 2013.

23. N. K. Denzin and Y. S. Lincoln, *The Sage Handbook of Qualitative Research*. Sage, 2011.
24. M. C. Taylor, *Critical Terms for Religious Studies*. University of Chicago Press, 2013.
25. H. Lune and B. L. Berg, *Qualitative Research Methods for the Social Sciences*. Pearson Education Limited, 2017.
26. V. Rabens, "The bible and ethics: Pathways for dialogue," *In die Skriflig*, vol. 51, no. 3, pp. 1–12, 2017.

Part II

Security Issues in the Metaverse

Chapter 7

A Review on Security Solutions for Virtual Reality Systems

Md Imtiaz Kamrul

7.1 INTRODUCTION

The Metaverse has drawn considerable attention due to its cutting-edge technological advancements, attracting interest from both industry and academia. It is a new transformation in technological advancements, providing a virtual space for individuals to experience a complex integration of 3D multi-user environments. The Metaverse including other digital technologies such as the Internet of Things (IoT), IoT for Medical, and augmented reality (AR) for seamless integration of virtual and physical presence, virtual reality (VR) for immersive experiences is widely utilized and has vast scope for development [1]. The metaverse, a collaborative virtual space that combines enhanced physical and digital reality, gives rise to novel modes of human engagement, financial exchanges, and social connections. The metaverse also increases its capacity to be a social media platform by using its immersive environment for more realistic experiences and also creates a space for the marketplace for virtual and physical goods [2], including non-fungible tokens (NFTs). The architecture of the metaverse is designed to integrate all of these features into one place. While promoting progress in multiple sectors, it presents notable difficulties, especially in terms of privacy and security. The data integration and intensiveness raise a security concern, hence highlighting the necessity of taking protective actions to secure individuals' information in this virtual universe.

Nowadays, a truth about the virtual space is that if you do not spend your money on a product, you become the product, which is also true in the metaverse [1]. Social media platforms are excellent examples of this statement because they provide a free service to millions of people who are aware of their preferences [2]. These platforms use cookie and internet footprint tracking capabilities to continuously analyze individual behavior in order to develop their business model [3]. Those data we leave behind can already reveal a lot about an individual's personality, choices, and thinking, which can be potentially harmful if the data is compromised by attackers in some way [4]. The predictive capabilities of different models are increasing rapidly, which could give us a sense of what might happen in the Metaverse.

DOI: 10.1201/9781032663609-9

The following subsections discuss the possible predictive capacity of the different applications introduced by the metaverse.

A) Individual profiling: Nowadays, virtual space converts all of its product users, whereas, in the metaverse, even small amounts of information will be useful. Virtual platforms are already a powerful magnet for gathering information from users, and the metaverse will be an exponentially more powerful tool for business owners [1]. It will be a unified space for everyone – such as readers, content creators, business owners, and students. The vast amount of information prompted a major concern on security and privacy. Several data breach incidents reported recently, such as social engineering attacks, are already reported in various sources [5, 6]. In the metaverse, it is assumed that the possible cyber-attacks will increase exponentially because it contains not only physical information but also biological information of individuals. Figure 7.1 illustrates a metaverse environment that distinguishes between AR, VR, and augmented virtuality, with these virtual spaces capable of profiling information [7]. This type of sensitive information is at risk if the providers fail to ensure proper secure measurement.

B) Individual Privacy: There are multiple aspects to consider when it comes to metaverse privacy. The research identified three types of concerns that are particularly important: personal, behavioral, and communication information [7]. For example, personal information on social media platforms is used for doxing, which is the spreading of private information that is insulting to the victim [8]. Because of the plethora of real-world information about individuals and characteristics, the metaverse can also be an ideal environment for doxing. In the real world, any form of harassment demands a close presence, whereas in the virtual space, physical presence is not required, making it more convenient for attackers. Several reports on proliferation online have been studied in this research, including raiding, shaming, cyberbullying, video call bombing, and so on [8]. These risks are undoubtedly amplified by the metaverse, which is designed for multi-space connections. Figure 7.2 depicts the metaverse's privacy-related properties as well as potential threats from attackers in considering the specific situation. This information is reported in the Microsoft Security Development Lifecycle and a literature survey on the mixed reality [9]. It is critical to ensure that data storage and flow in the metaverse cannot be tampered with or modified by attackers. Data should also be available only to the intended user after completing the necessary security check, which will protect against denial-of-service attacks. Authorization, authentication [1], and confidentiality can be attractive targets for attackers who need to be secure by implementing various security protocols.

This chapter discusses a fundamental issue concerning metaverse security and privacy, as well as multiple attack scenarios and solutions for general and specific attacks. This survey also addressed the major challenges in developing the metaverse, including potential attack scenarios

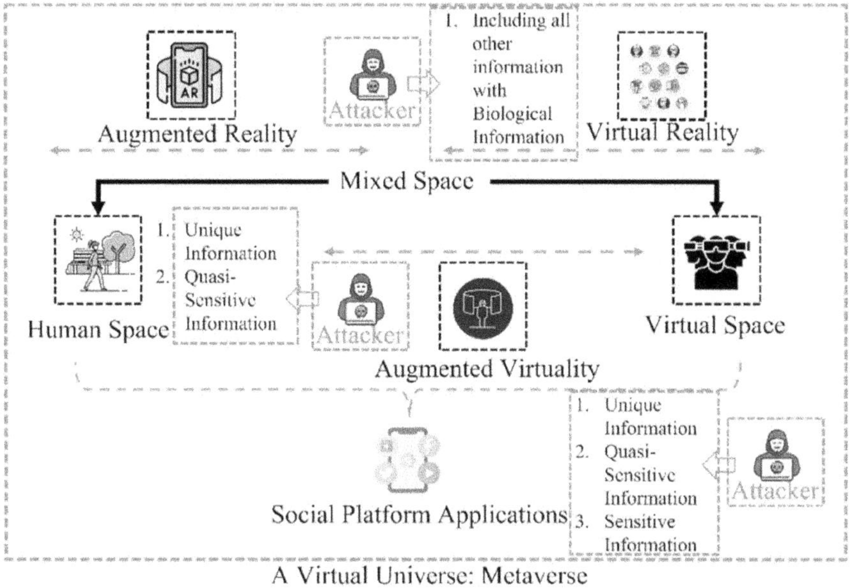

Figure 7.1 A metaverse environment with security concerns of individual profiling.

and vulnerable sections due to inadequate security measures. By discussing several possible metaverse attacks, as well as potential solutions, this chapter provides insightful insights and valuable guidelines for researchers and metaverse developers. Our contribution to this chapter has primarily four characteristics:

1) In-depth investigation and analysis of metaverse architecture, including potential vulnerabilities throughout the system. Also includes a specific example to help understand the potential attack scenario.
2) The survey on potential attack architecture, in brief, followed by a discussion of recent reported attacks. It also includes a framework for various types of attacks.
3) A possible solution has been proposed by providing critical insights into the attack types, such as a deep learning-based network intrusion detection system.
4) In addition, section IV includes an outline of potential and future research directions for creating an efficient metaverse environment.

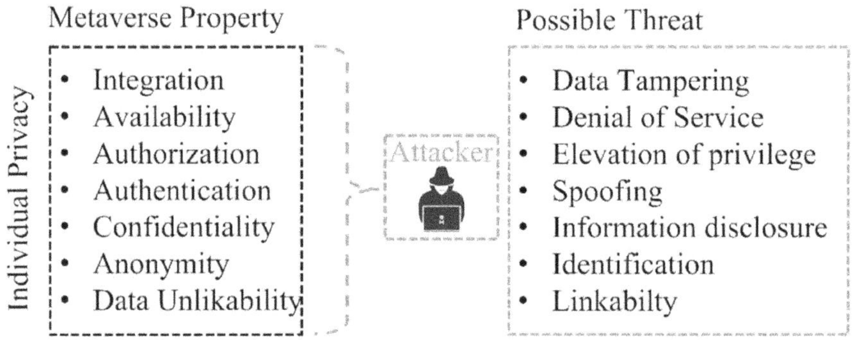

Figure 7.2 Privacy property of the Metaverse, including threats.

7.2 SECURITY CONCERNS AND CHALLENGES IN THE METAVERSE

7.2.1 MITM Attacks in the Metaverse

The 3D model reconstruction may be a target for attackers because it contains personal identification (PI) information [2], which allows users to have an enhanced experience in the metaverse or other virtual platforms. The general architecture of 3D model creation includes cloud database information handling that receives information from individual users. The 3D model is created on a separate server using sensitive personal information that is securely stored in a database to create a realistic avatar [2]. This information combines a variety of sensitive information, including retina patterns, fingerprint information, biological information, and so on. Figure 7.3 depicts a different attack scenario based on the communication phase with the users. MITM (man-in-the-middle) attacks occur during both the transmission and receiving phases. In the data transmission communication phase with the server, enable MITM to block or modify data traffic. Several types of attacks, including spoofing, false station-based, SSL/TLS [10], and side-channel attacks, have been reported here [11], all of which focus on the hardware vulnerabilities of a server or system. In the second scenario depicted in Figure 7.3, the hacker gains access to the 3D reconstructed model instead of the original user and collects data using social engineering, cross-site scripting, or a malicious web tool. This type of sensitive information could be used to extort or shame the victim online. In addition, the attacker can use the sensitive information to create a clone model of the individual identity in the virtual space, allowing him to replace the original person identity.

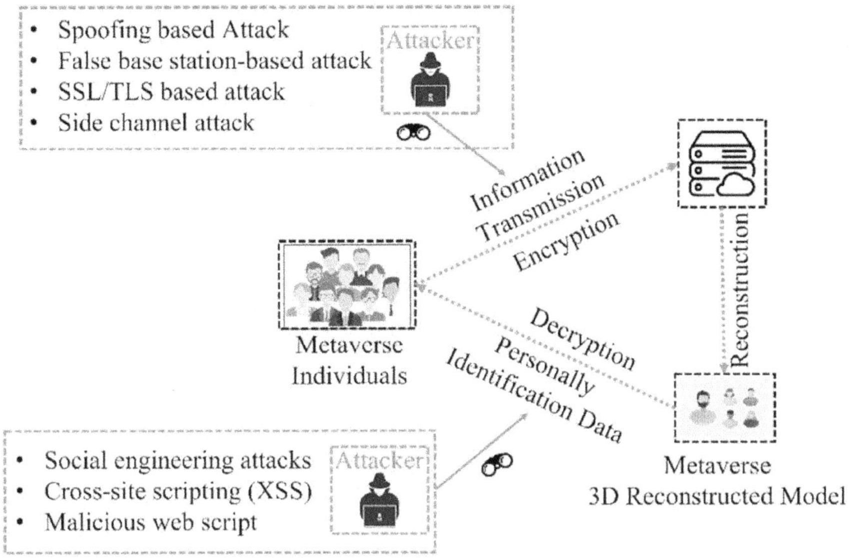

Figure 7.3 MITM attacks scenario to the 3D Reconstructed Avatar in the Metaverse.

7.2.2 Image Attacks in the Metaverse

The image data used to develop a privacy-conscious machine learning mode (PPML) could be a target for attackers. This study discusses the Jacobian-based saliency map attack pixel method [12]. Consider the following image, labeled (p,q), to describe the method. The saliency map observes each pixel's data and identifies the influenced pixel in image p using class q. PPML uses this assumption to correlate with the appropriate class. The saliency map is expressed by Equation 1 [12].

$$\text{Map}=\begin{cases} \frac{\partial f(p)_q}{\partial (p)_i} \cdot \sum_{q' \neq q} \frac{\partial f(p)_{q'}}{\partial (p)_i}, & \textit{if True} \\ 0\,, \quad \textit{Otherwise} & \end{cases} \tag{7.1}$$

where, *f(p)* represents the SoftMax probabilities and q' denotes the map's contradictory classes. The "true" equation satisfies the correlation condition, which states that pixel modifications increase the contradictory label. Attackers can also attack the PPML model using reversible up-sampling or down-sampling, but they require prior information to design a modified sub-sampling layer. As shown in Figure 7.4, the PPML model employs the saliency map to detect false labeling, restricted access, and denial of service.

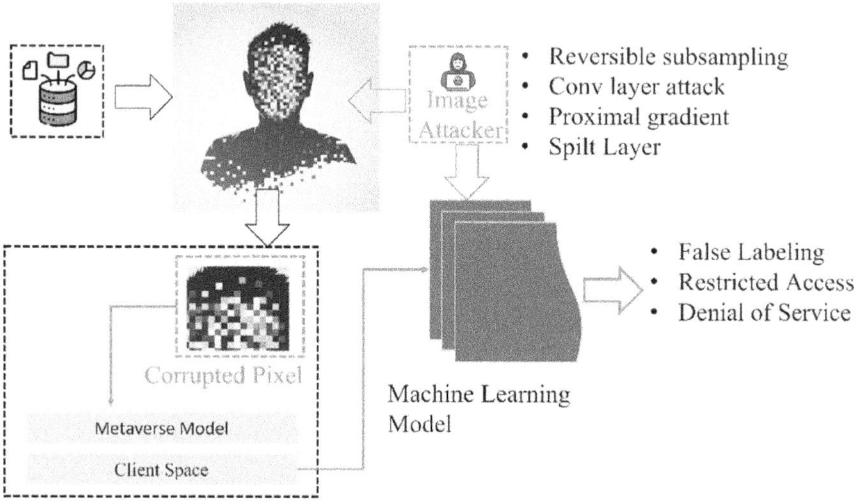

Figure 7.4 Pixel attacks by gradient learning network.

7.2.3 Neural Adversarial Attacks in the Metaverse

Deep learning models, including recurrent neural learning network models, were primarily developed to advance the Metaverse. However, the machine learning model may be vulnerable to adversarial attacks, such as Trojan virus attacks. Metaverse engineering cannot control a model's decision-making process, so it must be believed blindly. The most commonly attacked scenario reported by Trojan viruses in deep learning models involves targeting either physical features or imperceptible noise. Several traffic sign recognition benchmarks (GTSRB) are presented here [13], with invisible triggers used. Most of the mechanisms are currently being used primarily for clustering or group-basis testing, while the vast majority of the models only function as classification networks [14]. The various types of poisoned sample models are discussed here [15]. Another neural-based DNN model is used to identify the backdoor sample via reverse engineering. In addition, a competitive DRL method based on a separate training policy is proposed [15].

These types of backdoor attacks have the potential to harm automatic vehicles due to incorrect traffic sign recognition. As demonstrated in Figure 7.5, attackers can trigger data in different neural network layers without changing the model's features. The targeted model will train on the triggered data, resulting in a poisoned output without knowing the threats. Any kind of small perturbation in any stage of the neural machine learning model or injection of Trojan in any layer can be dangerous enough to manipulate the entire metaverse system.

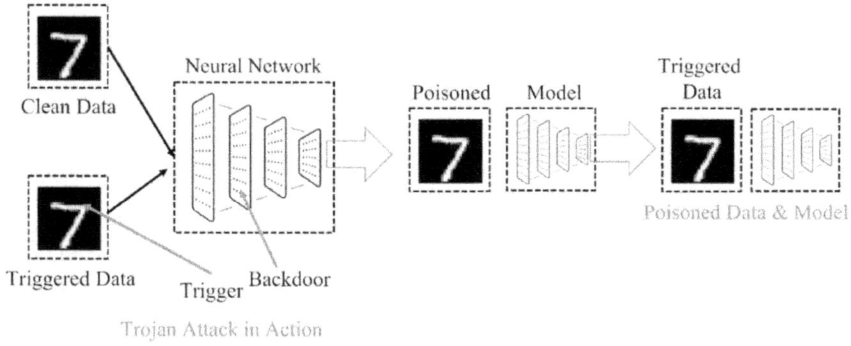

Figure 7.5 Illustration of neural Trojan attacks in the Metaverse.

7.2.4 Inference Attack in the Metaverse

The inference attack might significantly threaten the metaverse by exploiting machine learning model information, specifically about trained data. It could be a potential safety threat intended to expose the metaverse user's sensitive or personal identification information. Inference attacks on the metaverse involve several processes, including the creation of a shadow model, training the attack model, and evaluating the attack model [16]. In the first step, the attackers generally created a number of shadow models based on the information gathered in order to mimic the targeted model's behavioral approach. The attackers then use the shadow model to train the data of the targeted model, a technique known as a membership interference attack. They then evaluate the trained shadow model and use it to cause interference with the data used by the targeted model [16].

A membership interference attack, which is a query attack to determine whether or not a specific piece of information is available in a specific dataset, is discussed here. This type of anonymous attack on the metaverse is critical for revealing the user's identity. Another example was presented here, in which 1,120 input data scores were used to create a shadow model [16], and three key observations were made following training and membership interference attacks, namely the impact of batch size, the impact of how many rounds of training data occurred, and the impact of data augmentation effect.

Figure 7.6 shows an architectural view of the inference attack. The attackers create a large number of shadow models until they reach their destination. Once the shadow model recognizes a similarity with the targeted model, it attempts to mimic the targeted training data and behavioral features. As a result, it trains the targeted model on its shadow dataset without being detected. As a consequence, the entire ML model becomes vulnerable to attacks, posing a significant security risk to metaverse users.

There are several attack mechanisms explained here, all of which have certain distinctions and unique characteristics of the situation based on the field of application in the metaverse. It is challenging to address all issues and prevent threats in real time. However, security must be ensured in order to increase the credibility of the metaverse and encourage users. Otherwise, it is difficult to utilize the concept of metaverse in everyday circumstances. When people have proper security, they are less likely to lose their personal sensitive information. Table 7.1 summarizes all the threats discussed in this chapter for the specific application in the metaverse. Furthermore, the solutions to the attacks discussed here will be addressed in the following section using different solution approaches.

7.3 ADDRESSING METAVERSE ATTACKS

7.3.1 Network Intrusion Detection System

Network intrusion detection system (NIDS) architecture protects against web-based malicious attacks [17]. Traditional methods may be vulnerable to intrusion activity and unable to control large amounts of data, so it is preferable to use a modern machine learning model, such as deep learning, for the detection system. NIDS also uses a generative adversarial network (GAN) to generate lower-dimensional data, which increases flexibility and

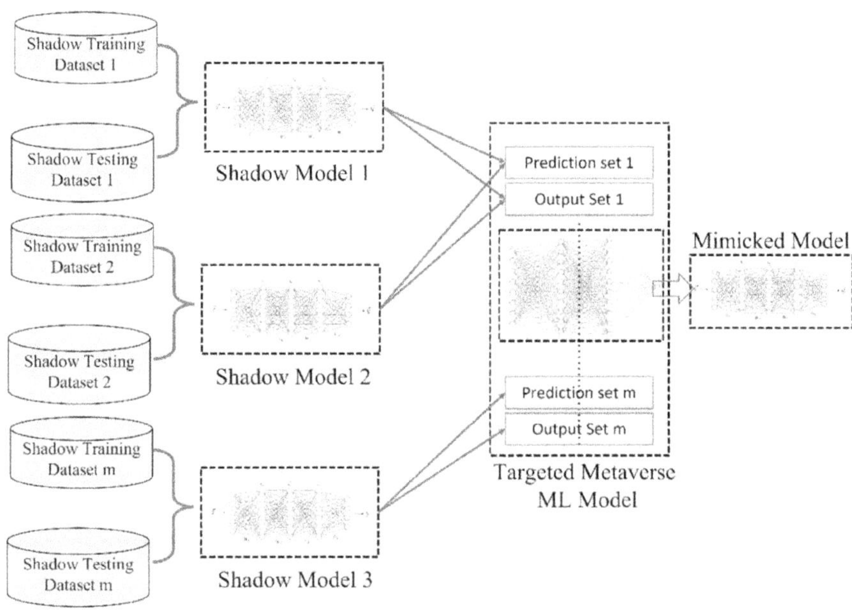

Figure 7.6 Workflow of the inference attack to the Metaverse ML model.

Table 7.1 Summary of the attack classification and threat to the Metaverse

Attack Type	*Threats*	*Application*
MITM attacks	Block or manipulate data traffic during transmission and receiving phases.	3D model reconstruction in the Metaverse (Spoofing, false station-based, SSL/TLS, and side-channel attacks.)
Image attacks	Jacobian-based saliency map attacks	Pixel data for privacy-conscious ML models (False labeling, Restricted access, Denial of service.)
Neural Adversarial attacks	Trojan virus base neural attacks affecting physical information and invisible noise adding to the information.	Deep learning models in the Metaverse (Incorrect traffic sign recognition in the metaverse or data manipulation in the metaverse)
Inference Attack	Exposes sensitive or personal identification information of	Machine learning model information in Metaverse (Membership interference attacks reveal user identity)

variability. In addition, a deep convolutional generative adversarial network (DCGAN) has recently been developed, capable of detecting complex, multidimensional cyber threats in the metaverse.

Figure 7.7 shows the deep learning-based NIDS architecture, which includes a preprocessing step to extract features using kernel principal component analysis (KPCA) [17] to improve the extraction of salient attack features. KCPA is very useful in distinguishing between various types of malicious attacks, including backdoor attacks, DoS attacks [18], MITM attacks, Trojan attacks, and XSS attacks [19]. It also has a convolutional neural network to detect the attack in the final stage. The overall deep learning-based NIDS system makes the detection model capable of identifying the attacks in the primary stage.

The early phase of the NIDS system uses the normalization process to regulate the IoT server data using a minimum-maximum scaling technique, as stated in equation 2 [17]. Following preliminary normalization steps, the feature extraction process will start with KCPA techniques, which help to reduce the data dimension.

$$Q_{normalization} = \frac{Q - \min(Q)}{\max(Q) - \min(Q)} \tag{7.2}$$

where Q represents the dataset's feature value in the range 0 to 1.

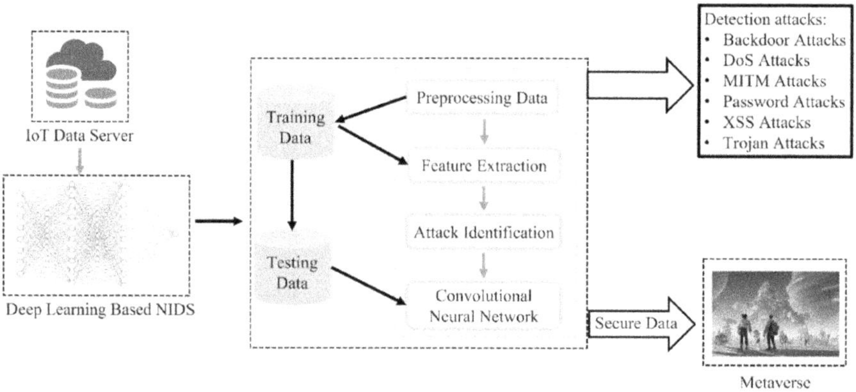

Figure 7.7 Deep learning-based NIDS architecture to secure Metaverse.

However, one of the main limitations of KCPA is that it does not deal with nonlinear data characteristics, as shown by the equation 3. The third approach for identifying attacks is to manually use the feature map obtained from the KCPA feature extraction process and forward it to the convolutional neural network.

$$Y : \varnothing \in X^a \rightarrow Y(\varnothing) \in R \tag{7.3}$$

Here, X represents the entire feature space within function Y.

The data management process of the convolutional kernel is followed by the following equation 4 [17]:

$$P_j^i = f\left(\sum_{l \in Kj} Q_l^{i-1} q M_{lj}^i + A_j^i\right) \tag{7.4}$$

where, P_j^i is the feature of layer map and M_{lj}^i is the function of CNN kernel.

The NIDS architecture's learning process goes in parallel and simultaneously enhances the database identification capacity. The structure is designed to detect different kinds of attacks, including backdoor attacks, DoS attacks, MITM attacks, XSS attacks, and Trojan attacks. The attack detection ability serves to ensure the confidentiality and safety of the metaverse.

7.3.2 Privacy Preservation and Advertising Strategy

Individual privacy is a top priority in the metaverse. The most promising aspect of the metaverse utilized augmented and VR technology, in which a metaverse service provider must allocate bandwidth for users to access the metaverse through an access point. In general, to increase their profit, metaverse service providers distribute several advertisements that contain the metaverse targeted user's marketing details, which could be an interesting source of information for attackers.

The privacy preservation and advertising strategy is a novel approach to ensuring the security of metaverse edge access points [20]. The initial phase involves properly allocating bandwidth and utilizing the covert communication method, which mixes noise with the transmitted information to protect the user's privacy. Because of this covert advising method [21], the architecture delivers targeted advertising to users while remaining undetected by attackers. Additionally, a meta-immersion metric is used in this architecture to improve individuals' engagement with potential security. Furthermore, this system incorporates jamming-access wireless access that can confuse potential listeners, thereby enhancing the efficiency of covert communications. Figure 7.8 provides the architectural view of this approach.

As shown in Figure 7.8, the metaverse service provider utilizes a covert communication strategy during bandwidth allocation to the metaverse user. The budget for the construction of virtual worlds is not an instantaneous measurement for the metaverse service provider; instead, it is measured over a period of time in the range of 0 to T1. The meta-immersion technology ensures user experience and satisfaction with the virtual world while providing high-quality data access. A jamming mechanism has also been established to prevent malicious data transmission.

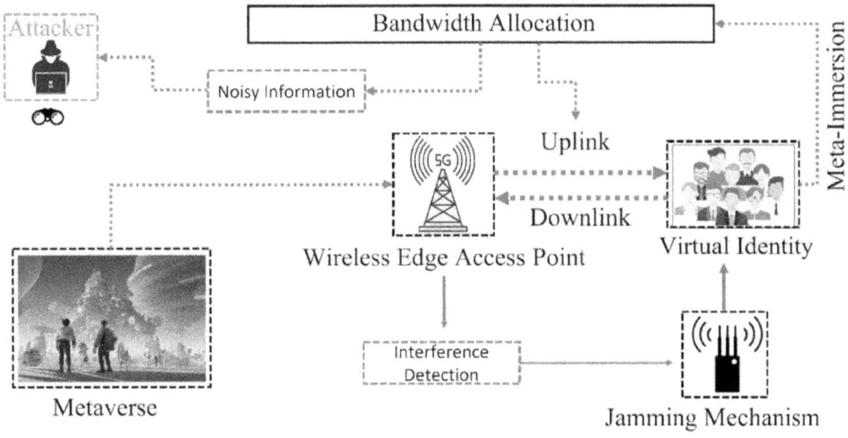

Figure 7.8 Architecture of privacy preservation and target advertising strategy.

7.3.3 Security Solution by Design Approaches

Another approach to ensuring security and taking precautions to prevent data loss is to design a security approach for every state of the metaverse [22]. Independent authority in every part of the metaverse offers an independent approach to security. Furthermore, a well-designed metaverse architecture with independent authority at each stage should ensure privacy by default, without any additional post-measurement. AI-based models may also be a viable solution for improving the efficiency of machine learning algorithms. To implement security in each stage, several tasks must be completed [22], such as (1) ensuring that the amount of personal data collected by the user is not distributed in each layer and that it is preserved after fulfilling the specific demand, (2) preserving personal data in a decentralized manner, (3) ensuring a processing level with higher aggregation throughput, (4) ensuring transparency in data management, and (5) ensuring a privacy policy in accordance with the law enforcement amendment.

7.3.4 XAI Security Measurement

XAI, or Explainable AI [23], is a promising intelligent solution for metaverse security. In this approach, the decision will be made based on AI judgment. The role of explainable AI is significant, particularly in three judgments [22] depicted in Figure 7.9. The XAI clarifies data prediction for specific data generated or handled by AI, which is essential to analyzing security measurements, preserving user privacy, and understanding AI workflow patterns. Furthermore, the XAI approach handles the privacy properties, allowing the machine learning model to process personal data while ensuring privacy during management in accordance with proper standards. All of these approaches enable XAI to detect potential threats, take precautions, and act as a virtual shield between the metaverse user and the service provider. The growing demand for XAI, as mentioned and reported in various research papers, highlights the importance of AI judgment in enhancing user trust.

The XAI implementation is required in the metaverse to provide trust and transparency of the system, which is critical in the metaverse due to human interaction with the AI-driven universe. Individuals can trust the system if AI decisions and judgments are explained, which helps to spread the technology. It will also improve privacy protection by processing user data in a specific manner, thereby defending against attacks via anomaly detection in the metaverse. XAI can explain the reason for potential security threats, allowing developers to improve policy levels and signature databases to take precautionary measures. Furthermore, XAI is an important device that should be used in the metaverse not only to protect the system's privacy but also to create an intelligent AI-powered verse that will enhance the trustworthiness of the virtual world.

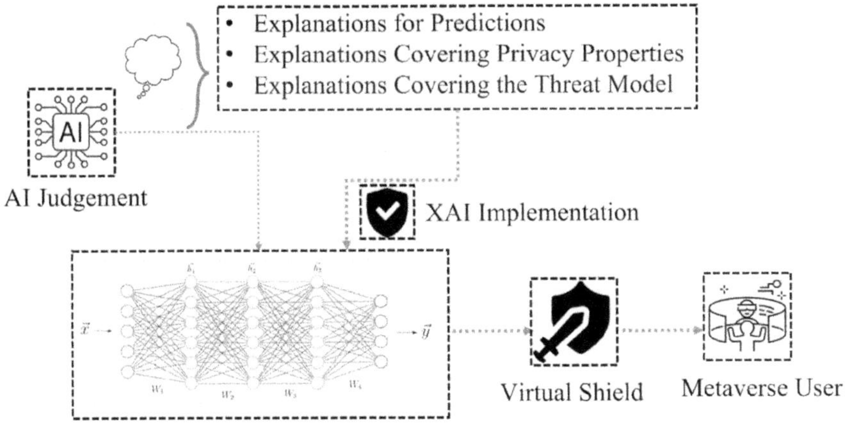

Figure 7.9 XAI implementation to provide security for metaverse users.

7.4 FUTURE DIRECTION AND DISCUSSION

The purpose of this chapter is to discuss and analyze the current issue in the metaverse regarding security and privacy concerns, based on a recently published research paper. This study discussed potential security threats, including a few specific attacks. Typically, the metaverse is still in development, and researchers are working to develop it; thus, various research papers have revealed many aspects of the field and must be considered in future work. We aim to summarize the most vulnerable and risky attacks on the metaverse so that others can move forward with further development. As previously stated, privacy and security concerns in the metaverse are of great interest to researchers, and AI-based security, specifically neural-based deep learning security systems, are the most important security architectures.

In this chapter, we discussed MITM attacks, which attackers commonly use to gather personal information such as sensitive credit card or financial transaction information through spoofing, false station-based, SSL/TLS, and side-channel attacks. We also addressed the risks that an individual may face as a result of data leakage, as well as the possible aspects of the metaverse and how the metaverse could be a potential target for adversarial attacks. Metaverse is an immersive virtual field that is responsible not only for providing an engaging experience to users but also for providing a flexible and robust security solution by default so that individuals are not concerned about their privacy.

Several security solution approaches, particularly those that take into account the vast extent of the metaverse, have also been discussed here. The metaverse is unlike other types of virtual spaces in that it combines all VR technology, including its immersive features, to make it more realistic. On the one hand, it creates space for the interactive battlefield to work while also increasing the number of attack possibilities. It is necessary to acquire the best security architecture in order to provide and ensure the safety of the metaverse. This survey chapter addressed the Privacy Preservation Approach for Metaverse Service Providers, which allows for safety through the use of covert communication techniques. On the other hand, NIDS architecture is very beneficial in detecting complex, multidimensional cyber threats in the metaverse. It is also crucial to provide security at each stage of the metaverse, and this must be done in a decentralized manner, as discussed in our third section on security solutions. Overall, we summarized the most useful architecture to solve the metaverse issue, which will be a valuable resource for future work.

REFERENCES

1. R. D. Pietro and S. Cresci, "Metaverse: Security and Privacy Issues," in *2021 Third IEEE International Conference on Trust, Privacy and Security in Intelligent Systems and Applications (TPS-ISA)*, 13–15 December 2021, pp. 281–288, doi: 10.1109/TPSISA52974.2021.00032.
2. I. Vladimirov, M. Nenova, D. Nikolova, and Z. Terneva, "Security and Privacy Protection Obstacles with 3D Reconstructed Models of People in Applications and the Metaverse: A Survey," in *2022 57th International Scientific Conference on Information, Communication and Energy Systems and Technologies (ICEST)*, 16–18 June 2022, pp. 1–4, doi: 10.1109/ICEST55168.2022.9828791.
3. P. Laperdrix, N. Bielova, B. Baudry, and G. Avoine, "Browser Fingerprinting: A Survey," *ACM Transactions on the Web*, vol. 14, no. 2, p. Article 8, 2020, doi: 10.1145/3386040.
4. S. Sahmim and H. Gharsellaoui, "Privacy and Security in Internet-based Computing: Cloud Computing, Internet of Things, Cloud of Things: a review," *Procedia Computer Science*, vol. 112, pp. 1516–1522, 1 January 2017, doi: https://doi.org/10.1016/j.procs.2017.08.050.
5. M. Kosinski, D. Stillwell, and T. Graepel, "Private Traits and Attributes Are Predictable from Digital Records of Human Behavior," *Proceedings of the National Academy of Sciences*, vol. 110, no. 15, pp. 5802–5805, 2013, doi: doi:10.1073/pnas.1218772110.
6. M. D. Conover, B. Goncalves, J. Ratkiewicz, A. Flammini, and F. Menczer, "Predicting the Political Alignment of Twitter Users," in *2011 IEEE Third International Conference on Privacy, Security, Risk and Trust and 2011 IEEE Third International Conference on Social Computing*, 9–11 October 2011, pp. 192–199, doi: 10.1109/PASSAT/SocialCom.2011.34.
7. B. Falchuk, S. Loeb, and R. Neff, "The Social Metaverse: Battle for Privacy," *IEEE Technology and Society Magazine*, vol. 37, no. 2, pp. 52–61, 2018, doi: 10.1109/MTS.2018.2826060.

8. P. Snyder, P. Doerfler, C. Kanich, and D. McCoy, "Fifteen Minutes of Unwanted Fame: Detecting and Characterizing Doxing," presented at the Proceedings of the 2017 Internet Measurement Conference, London, United Kingdom, 2017. Available: https://doi.org/10.1145/3131365.3131385.
9. J. A. D. Guzman, K. Thilakarathna, and A. Seneviratne, "Security and Privacy Approaches in Mixed Reality: A Literature Survey," *ACM Computing Surveys,* vol. 52, no. 6, p. Article 110, 2019, doi: 10.1145/3359626.
10. M. Conti, N. Dragoni, and V. Lesyk, "A Survey of Man in the Middle Attacks," *IEEE Communications Surveys & Tutorials,* vol. 18, no. 3, pp. 2027–2051, 2016, doi: 10.1109/COMST.2016.2548426.
11. I. Vladimirov, D. Nikolova, and Z. Terneva, "Hardware Implementation and Comparison of CRA and TRA when Trying to Recover the AES-128 Key," in *2020 55th International Scientific Conference on Information, Communication and Energy Systems and Technologies (ICEST)*, 10–12 September 2020, pp. 256–258, doi: 10.1109/ICEST49890.2020.9232709.
12. S. A. Khowaja, I. H. Lee, K. Dev, M. A. Jarwar, and N. M. F. Qureshi, "Get Your Foes Fooled: Proximal Gradient Split Learning for Defense Against Model Inversion Attacks on IoMT Data," *IEEE Transactions on Network Science and Engineering,* vol. 10, no. 5, pp. 2607–2616, 2023, doi: 10.1109/TNSE.2022.3188575.
13. Y. Li, S. Zhang, W. Wang, and H. Song, "Backdoor Attacks to Deep Learning Models and Countermeasures: A Survey," *IEEE Open Journal of the Computer Society,* vol. 4, no. 01, pp. 134–146, 2023, doi: 10.1109/OJCS.2023.3267221.
14. J. L. B. Tran, and A. Madry, "Spectral Signatures in Backdoor Attacks," *NeurIPS Proceedings,* vol. 31, pp. 1–11, 2018.
15. Y. Yu, J. Liu, H. Guo, B. Mao, and N. Kato, "A Spatiotemporal Backdoor Attack Against Behavior-Oriented Decision Makers in Metaverse: From Perspective of Autonomous Driving," *IEEE Journal on Selected Areas in Communications,* vol. 42, no. 4, pp. 948–962, 2024, doi: 10.1109/JSAC.2023.3345379.
16. R. Shokri, M. Stronati, C. Song, and V. Shmatikov, "Membership Inference Attacks Against Machine Learning Models," presented at the 2017 IEEE Symposium on Security and Privacy (SP), 2017. [Online]. Available: https://doi.ieeecomputersociety.org/10.1109/SP.2017.41.
17. T. Gaber, J. B. Awotunde, M. Torky, S. A. Ajagbe, M. Hammoudeh, and W. Li, "Metaverse-IDS: Deep Learning-Based Intrusion Detection System for Metaverse-IoT Networks," *Internet of Things*, vol. 24, p. 100977, 2023/12/01/ 2023, doi: https://doi.org/10.1016/j.iot.2023.100977.
18. Z. Alkhalil, C. Hewage, L. Nawaf, and I. Khan, "Phishing Attacks: A Recent Comprehensive Study and a New Anatomy," (in English), *Frontiers in Computer Science, Review*, vol. 3, 9 March 2021, doi: 10.3389/fcomp.2021.563060.
19. G. E. Rodríguez, J. G. Torres, P. Flores, and D. E. Benavides, "Cross-site Scripting (XSS) Attacks and Mitigation: A Survey," *Computer Networks,* vol. 166, p. 106960, 15 January 2020, doi: https://doi.org/10.1016/j.comnet.2019.106960.

20. H. Du, D. Niyato, C. Miao, J. Kang, and D. I. Kim, "Optimal Targeted Advertising Strategy for Secure Wireless Edge Metaverse," in *GLOBECOM 2022–2022 IEEE Global Communications Conference*, 4–8 December 2022, pp. 4346–4351, doi: 10.1109/GLOBECOM48099.2022.10001331.
21. I. Ullah, R. Boreli, and S. S. Kanhere, "Privacy in Targeted Advertising on Mobile Devices: A Survey," *International Journal of Information Security,* vol. 22, no. 3, pp. 647–678, 1 June 2023, doi: 10.1007/s10207-022-00655-x.
22. C. Sandeepa, S. Wang, and M. Liyanage, "Privacy of the Metaverse: Current Issues, AI Attacks, and Possible Solutions," in *2023 IEEE International Conference on Metaverse Computing, Networking and Applications (MetaCom)*, 26–28 June 2023, pp. 234–241, doi: 10.1109/MetaCom57706.2023.00052.
23. A. Kuppa and N. A. Le-Khac, "Black Box Attacks on Explainable Artificial Intelligence(XAI) methods in Cyber Security," in *2020 International Joint Conference on Neural Networks (IJCNN)*, 19–24 July 2020, pp. 1–8, doi: 10.1109/IJCNN48605.2020.9206780.

Chapter 8

Metaverse Security

Principles and Experiments

Md Rafi Islam and Shuhui Li

8.1 INTRODUCTION

The Metaverse represents a transformative frontier in digital innovation, blending physical and virtual realities through technologies such as virtual reality (VR), augmented reality (AR), artificial intelligence (AI), and blockchain. Envisioned as a seamless, immersive environment for interaction, the Metaverse transcends traditional digital ecosystems by offering real-time 3D spaces for social engagement, commerce, education, and entertainment. However, as the Metaverse grows in scope and adoption, it introduces complex security challenges, ranging from safeguarding user data to ensuring secure interoperability among diverse platforms. With personal data increasingly embedded in virtual interactions, protecting users' privacy and identity in the Metaverse has become a pressing concern. Moreover, the reliance on IoT-connected devices and decentralized infrastructures exposes the ecosystem to cyber threats that demand innovative mitigation strategies. This chapter explores foundational principles of Metaverse security and the experiments conducted to address critical vulnerabilities, highlighting advancements in secure communication, privacy protection, and cyber resilience.

8.2 BACKGROUND

The Metaverse, an evolving digital ecosystem blending physical and virtual realities, presents unique opportunities and complex challenges. The integration of VR, AR, and AI on a single platform requires strict security measures to address its vulnerabilities.

One critical challenge is privacy. The Metaverse extensively collects and processes sensitive user data, including behavioral, biometric, and interaction details. Research shows that adversarial VR environments can covertly extract distinct data attributes, such as physical dimensions and demographic information, through subtle cues in user interactions [1, 2]. These concerns highlight the potential misuse of sensitive data by malicious

DOI: 10.1201/9781032663609-10

actors. Moreover, the integration of IoT devices into Metaverse applications introduces a vast attack surface, further exacerbating the risks of data breaches and unauthorized surveillance [3, 4].

Interoperability is another pressing concern, as Metaverse platforms are inherently decentralized yet lack standardized protocols for secure interactions. The secure web objects (SWO) framework proposes a decentralized, data-centric approach to address these limitations, emphasizing secure object-level interoperability without relying on centralized infrastructures [5, 6]. This framework aims to mitigate the inefficiencies of traditional client-server communication models and enable seamless cross-platform interactions.

Cyber threat detection remains a significant challenge, particularly in educational and vehicular Metaverse applications. Emerging cyber threats, such as ransomware, session hijacking, and network intrusions, pose significant risks. Explainable AI (XAI)-based intrusion detection systems, using techniques such as Shapley Additive Explanations (SHAP), Local Interpretable Model-agnostic Explanation (LIME), and bi-layer Bi-directional Long Short-Term Memory (BiLSTM), aim to enhance the interpretability and trustworthiness of security measures [2, 3]. However, improving transparency and reliability in these models is an ongoing endeavor. Additionally, reputation-based mechanisms have been proposed to combat malicious behavior in IoT-enabled vehicular Metaverse systems, ensuring data integrity and maintaining the quality of experience for users [4].

The vulnerabilities of VR devices present yet another critical challenge. Weaknesses in firmware can lead to user authentication flaws and open doors for hijacking scenarios. Different methods have been developed to identify and address these vulnerabilities, but the rapid evolution of VR technology necessitates continuous research to stay ahead of potential exploits [2, 4].

8.3 SCHEME CATEGORY

In the subsequent sections, we will delve into three distinct scheme categories, each addressing a critical facet of Metaverse security. Scheme Category 1 will investigate how adversarial VR designs leverage telemetry data to infer user attributes. Scheme Category 2 will describe the integration of machine learning models and intuitive mechanisms to enhance security and user trust in virtual environments. Finally, Scheme Category 3 will discuss solutions leveraging blockchain-based decentralized identity management, cryptographic SWOs for interoperable interactions, and reputation-based frameworks to ensure data integrity and mitigate risks in IoT-enabled vehicular systems. These schemes collectively address the multifaceted challenges of Metaverse security, offering a holistic perspective on protecting this evolving digital ecosystem.

8.3.1 Scheme Category 1: Privacy and Vulnerability in Immersive Environments

The immersive nature of the Metaverse introduces unique privacy risks and vulnerabilities, particularly within VR systems. This category focuses on innovative approaches to safeguard sensitive user attributes and ensure secure interactions in VR environments. From detecting adversarial designs that exploit telemetry data to infer demographic and cognitive traits, to developing non-intrusive security indicators for secure transitions, these solutions address the foundational need for trust and safety in immersive digital spaces.

Unlike conventional digital platforms, VR systems generate continuous telemetry data streams encompassing motion patterns, audio inputs, and spatial interactions [1]. These data streams are critical for creating seamless and immersive experiences. However, as the authors in [1] emphasize, they also present significant privacy vulnerabilities, particularly when leveraged through adversarial designs that exploit the unique data-rich environment of VR. Adversarial VR designs use gamified elements to covertly elicit sensitive user information under the guise of benign interactions. These designs exploit VR's engaging and immersive nature to extract anthropometric details, behavioral patterns, cognitive metrics, and demographic attributes. Article [1] examines experimental findings from an adversarial VR application, highlighting its framework, results, and implications for privacy, ethics, and regulation.

Participants in the study interacted with puzzles in an escape room-style virtual environment designed to covertly harvest user attributes during gameplay. Figure 8.1 illustrates the threat model in VR with different types of attackers. Fifty participants from diverse demographic backgrounds used consumer-grade VR devices, such as the HTC Vive, Vive Pro 2, and Oculus Quest 2. These devices captured telemetry data, including motion patterns, reaction times, and spatial interactions, which were analyzed to infer sensitive attributes. The virtual environment comprised puzzle-filled rooms, each eliciting specific user data. Tasks required participants to perform movements such as stretching or stepping back, indirectly revealing anthropometric attributes such as height, wingspan, and physical play area dimensions. Ground truth data, including participants' anthropometrics and cognitive baselines, were collected through surveys and direct measurements for validation. Additionally, cognitive tasks evaluated reaction times and problem-solving abilities, while exploratory challenges analyzed spatial movements to infer room dimensions and geolocation. Non-instrumental tasks masked the data collection intentions, distracting participants from

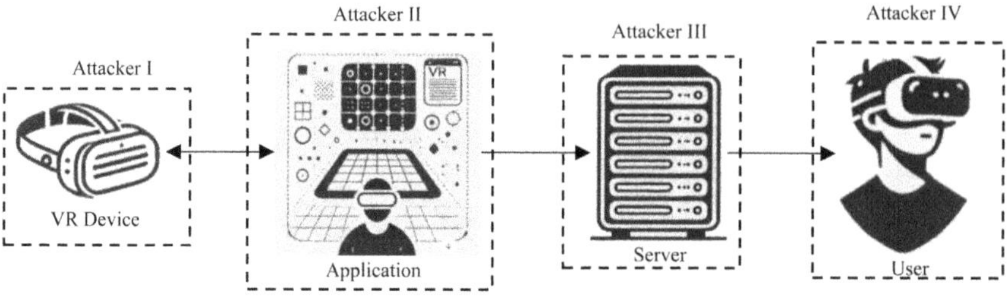

Figure 8.1 Virtual reality threat model.

the targeted harvesting mechanisms. High-frequency telemetry data streams were analyzed using statistical models to evaluate the precision of adversarial inferences.

The results demonstrated the effectiveness of adversarial VR designs in covertly extracting a wide range of user attributes. Biometric data collection achieved high accuracy, with continuous motion tracking measuring height within 5 cm for 76% of participants ($R^2 = 0.75$) and wingspan within 12 cm for 98% ($R^2 = 0.68$). Behavioral cues, such as button-pressing patterns, identified handedness with 96% accuracy and arm dominance with 100% accuracy in cases of significant asymmetry. Gamified tasks detected visual impairments, such as nearsightedness and color blindness, with accuracies exceeding 80% and 100%, respectively. Environmental attributes were also effectively captured. Spatial tracking inferred physical play area dimensions with 96% accuracy within 3 m^2, and network latency analysis pinpointed participant locations within 500 km for 94% of cases. Telemetry data further revealed socioeconomic indicators by inferring VR device specifications, such as refresh rates (100% accuracy ±3 Hz) and resolution and field of view (over 95% accuracy). Cognitive assessment tasks embedded within the puzzles extracted insights into participants' traits and behaviors. For example, the Adapted Montreal Cognitive Assessment (MoCA) tasks covertly evaluated cognitive acuity with 94% accuracy, while reaction time analysis inferred demographic traits, such as age, with 88% accuracy. Interaction patterns also revealed cognitive traits such as attention span and language proficiency. The article inferred demographic attributes and uniquely identified users by combining biometric, environmental, and behavioral data. Telemetry analysis determined gender with 98% accuracy, estimated age within 1.5 years for all participants, and classified ethnicity with 98% accuracy. Aggregated data streams enabled near-perfect re-identification of participants across sessions, demonstrating the potential for cross-platform tracking.

The findings underscore the profound privacy risks posed by adversarial VR designs. By leveraging telemetry data and interactive elements, adversaries can infer a wide spectrum of user attributes with alarming precision.

These designs actively guide users into behaviors that reveal specific attributes, often without their knowledge. The use of machine learning amplifies the scope and accuracy of these inferences, enabling detailed profiling. Unlike passive data collection, adversarial mechanisms exploit VR's interactive nature, covertly gathering data through movements and reactions. As VR becomes integral to domains such as education, healthcare, and entertainment, the privacy risks associated with adversarial designs grow exponentially. The creation of comprehensive user profiles raises concerns about surveillance, discrimination, and manipulative targeting. These risks are particularly troubling in sensitive applications such as therapy and remote work. The covert nature of adversarial designs challenges existing legal frameworks, such as the GDPR and CCPA, which primarily regulate overt data collection [1]. Current consent mechanisms in VR are inadequate, as users unknowingly reveal sensitive information through natural interactions. Ethical guidelines must prioritize transparency, requiring developers to disclose data collection practices and provide user-friendly privacy controls.

Despite its contributions, article [1] highlights limitations that warrant further investigation. The experimental setup focused on consumer-grade devices, leaving hardware-level threats and emerging technologies, such as eye tracking, unexplored. Additionally, the demographic composition of participants may limit the generalizability of findings. Future research should explore countermeasures, such as differential privacy techniques, and examine the long-term societal impacts of VR privacy breaches.

8.3.2 Scheme Category 2: Machine Learning Model-Driven Security and Intuitive Safeguards in Virtual Environments

The rapid expansion of virtual environments, particularly within the Metaverse, necessitates sophisticated security frameworks that not only detect and mitigate threats but also enhance user experience. This scheme category highlights the intersection of advanced machine learning models and intuitive safeguards to address emerging security challenges. It examines the application of deep neural networks (DNNs) for robust network intrusion detection, ensuring real-time protection against cyberattacks. Additionally, a BiLSTM-based framework is explored to identify firmware vulnerabilities in VR devices, reinforcing the security of foundational Metaverse technologies. Complementing these technical solutions, intuitive VR security indicators, such as visual and haptic signals, are evaluated to provide users with non-intrusive mechanisms for signaling secure transitions. Together, these approaches exemplify the integration of AI-driven models and user-focused safeguards, creating a comprehensive security paradigm for virtual environments. Figure 8.2 gives an overview of the Metaverse cyber threat detection model.

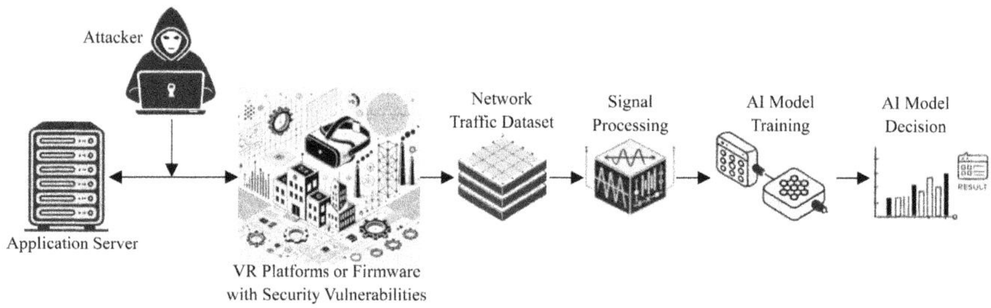

Figure 8.2 Metaverse cyber threat detection illustration.

These platforms in the Metaverse rely on IoT systems that introduce vulnerabilities, necessitating advanced measures to ensure security and trust. Article [3] presents a DNN-based network intrusion detection system (NIDS) with robust explainability tailored to the unique security demands of the Metaverse, bridging critical security gaps in the field.

The proposed system in [3] enhances the security of Metaverse learning platforms by detecting a variety of cyber threats, including DDoS attacks, malware, and injection attacks. The system emphasizes scalability, interpretability, and efficiency. DNN architecture was chosen for its capacity for automatic feature learning and high accuracy. It employs ReLU activation functions in hidden layers, with Softmax activation for multi-class outputs. Regularization techniques such as L1L2 were used to prevent overfitting, and the Adam optimizer ensures rapid convergence during training. The model supports both binary (normal vs. anomalous) and multi-class classifications, achieving robust predictive accuracy across diverse attack scenarios. The system leverages three recent IoT cybersecurity datasets – EdgeIIoT, CICIoT2023, and UNSW-NB15 – which include real-world cyberattack scenarios and IoT-specific features, providing a comprehensive basis for training and evaluation. Data preprocessing involves removing redundant features, such as source IP, and normalizing data to eliminate biases. Synthetic Minority Oversampling Technique (SMOTE) was employed to balance the dataset classes, ensuring the model is trained effectively. Feature selection is optimized using a decision tree-based method that ranks features by relevance. A threshold is applied to retain the top 30 features, reducing computational complexity while maintaining classification accuracy. Features are extracted through network traffic analysis tools such as Zeek and stored in CSV format for processing. The system incorporates SHAP and LIME techniques to enhance trust and usability. SHAP values can be calculated with Equation (8.1):

$$\phi_i(f,x) = \sum_{z' \subseteq x'} \frac{|z'|!\left(M - |z'| - 1\right)!}{M!}\left[f_x(z') - f_x\left(z' - f_x(z')\right)i\right]. \tag{8.1}$$

Here, ϕi denotes the Shapley value for feature i, M is the total number of features in model f with input data points x, z', and (z')i represents the model output with and without the target feature.

The LIME interpretability constraint can be expressed as follows:

$$explanation(x) = \arg\min L_{g \in G}(f, g, \pi_x) + ©(g). \tag{8.2}$$

Here, an explanation model, represented as x, is the model g that minimizes the loss L , reflecting how closely it aligns with the predictions of the original model f (such as a DNN) while ensuring minimal model complexity. G encompasses all potential explanations, while the proximity measures π_x define the neighborhood size around instance x considered for the explanation.

SHAP provides a global overview of feature importance, offering insights into the model's decision-making processes across the dataset. LIME focuses on individual predictions, identifying the features most critical to specific decisions. Together, these methods ensure the system is interpretable and actionable for cybersecurity professionals, fostering trust in the NIDS.

The system's performance is evaluated using metrics such as accuracy, F1-score, precision, and recall, demonstrating its versatility in both binary and multi-class classification tasks. The DNN achieves an F1-score of 99.9% in binary classification (normal vs. anomalous) and 71.22% in multi-class classification. Detection rates for certain attack types, such as DDoS TCP and ICMP, are high, while others, such as SQL Injection, are less effective due to data imbalance. Using the CICIoT2023 dataset, the system achieves 99.1% accuracy in binary classification. On the UNSW-NB15 dataset, it delivers multi-class accuracy of 84.74%, comparable to state-of-the-art models, while excelling in binary predictions with a precision of 99.9%.

The article employs Confidence Impact Ratio (CIR) and Decision Impact Ratio (DIR) to quantify explainability. SHAP outperforms LIME in both metrics, demonstrating its reliability in explaining predictions. Visualizations generated by SHAP and LIME provide actionable insights. For example, SHAP identifies the importance of TCP destination ports in detecting DDoS attacks, while LIME validates the model's confidence in individual predictions. These insights guide cybersecurity experts in understanding and mitigating potential threats, reinforcing the system's interpretability and practical utility. Future research could investigate the computational complexities of SHAP and LIME XAI methods while exploring the integration of an explainable adversarial protection mechanism to defend the NIDS from DNN adversarial attacks. Keeping the emphasis on security and explainability, the next article [2] shifts focus to VR as a core technology in the Metaverse, highlighting its transformative potential and inherent vulnerabilities.

Previous studies have emphasized the security risks inherent to the Metaverse, including identity theft and data breaches within VR systems [2]. Robust authentication protocols, particularly those utilizing biometric and blockchain-based solutions, have been explored to mitigate these threats. Privacy protection has also been a focus, with techniques such as multi-authority schemes designed to safeguard user data. However, attack vectors such as man-in-the-room exploits and de-anonymization continue to exploit vulnerabilities in VR platforms, underscoring the need for advanced detection and mitigation tools. Despite these advancements, research addressing firmware-specific vulnerabilities in VR devices remains scarce.

To address risks such as firmware weaknesses that can lead to data breaches or system manipulation, VRVul-Discovery, a BiLSTM-based vulnerability discovery framework, is introduced in [2]. The framework identifies known and previously undocumented vulnerabilities with high precision by analyzing critical attributes such as VR device startup protocols, authentication methods, and scenario interactions. Experiments validating VRVul-Discovery reveal its effectiveness in uncovering seven previously undocumented vulnerabilities, marking a significant step in enhancing the security of VR-enabled Metaverse interactions.

VRVul-Discovery employs a bi-layer BiLSTM network to identify vulnerabilities in VR firmware and determine their implications for hijackable scenarios [2]. The deep learning model analyzes the relationships between multidimensional attributes, enabling precise identification and mitigation strategies for vulnerabilities. The framework processes a dataset comprising both known and unknown vulnerabilities. The attributes of VR firmware analyzed in article [2] are categorized into three key areas: startup processes, user authentication mechanisms, and application scenario interactions. Firmware attributes encompass functional categories (e.g., gaming or browsing), protection mechanisms such as encryption and serial number checks, open interfaces (e.g., network or body-based inputs), third-party library paths, and known vulnerability statuses. These attributes define how firmware operates during initialization, revealing potential exploitation points. Authentication attributes include protocol interfaces, biometric or social authentication factors, authentication frequencies, and session storage methods. These dimensions are critical for ensuring resilient user authentication. Interaction attributes focus on body movements (e.g., inertial and optical sensing), biometric inputs (e.g., facial recognition), digital data (e.g., social and account information), and physical interactions with the environment. Together, these elements define secure and user-friendly application settings. Attributes undergo normalization to standardize data, ensuring compatibility with the model. The bi-layer BiLSTM network uses forward and backward LSTM units to analyze temporal and contextual relationships, while dropout layers prevent overfitting, enhancing model generalizability. Final outputs are refined through fully connected layers and Softmax functions to classify firmware vulnerabilities probabilistically. The iterative training process, which incorporates extensive data normalization and feedback adjustments, ensures high accuracy in detecting firmware vulnerabilities.

Once vulnerabilities are identified, the framework employs a hijackable scenario decision model to map vulnerabilities to specific application scenarios. Attributes from startup, authentication, and interaction dimensions are normalized and analyzed by the bi-layer BiLSTM network. The model predicts hijackable scenario attributes with high precision and verifies predictions using external tools to refine results. This process identifies specific scenarios vulnerable to hijacking, enabling targeted mitigation strategies.

The VRVul-Discovery system was validated through comprehensive experiments using a dataset of 1,872 known vulnerability firmware files, 515 potentially exploitable files, and 2,247 third-party application firmware samples. Feature extraction was conducted using tools such as IDA and WinHex, enabling multidimensional data analysis. Training and evaluation combined these datasets to ensure robust vulnerability identification and scenario mapping. The model underwent 180 training iterations, achieving an accuracy of 85.89% and reducing loss to 13.26%. By analyzing 400 firmware samples, the model identified 16 with high probabilities of vulnerability, demonstrating its capacity to uncover previously unknown risks. The hijackable scenario model achieved predictive accuracies exceeding 79% across attribute categories. Network interfaces and authentication factors showed the highest accuracy, demonstrating the model's ability to distinguish vulnerable scenarios effectively. Verification of identified vulnerabilities resulted in the discovery of seven previously unknown vulnerabilities, each validated and documented with formal vulnerability IDs. Future research could enhance the model by incorporating user behavior patterns, environmental contexts, and application-specific functionalities to better detect vulnerabilities in VR application scenarios.

The success of the VRVul-Discovery system in uncovering previously unknown vulnerabilities underscores the critical need for robust frameworks to address security risks in VR devices and their firmware. As the Metaverse continues to evolve into a more interconnected and dynamic ecosystem, the complexity of its infrastructure expands, bringing with it new challenges in safeguarding user data and ensuring secure interactions. This evolution is further accelerated by the efforts of major tech companies such as Meta, Microsoft, and Niantic, which are at the forefront of efforts to create interconnected sub-Metaverses akin to the World Wide Web within virtual spaces [6]. While these developments introduce innovative user experiences, they pose significant security risks, including identity exposure and data breaches. Addressing these challenges requires mechanisms to signal secure transitions between sub-Metaverses, analogous to the padlock icon in web browsers that indicates encryption. To fill this gap, article [6] designed and evaluated security indicators tailored for the Metaverse, employing participatory design and user studies to ensure their effectiveness, intuitiveness, and non-intrusiveness.

The article investigates key questions to understand VR security indicators: the general design dimensions for such indicators, their effectiveness in signaling secure transitions, usability, balance between noticeability and non-intrusiveness, and alignment with user preferences for daily VR interactions. The authors conducted semi-structured interviews with eight VR and security experts to address these questions. Thematic analysis revealed two primary dimensions influencing security indicator design: the design space and the context space. The design space comprises key aspects such as signal type (visual, auditory, or haptic), timing (constant or risk-based), placement (in the field of view or near portals), visual representation (minimalistic shapes or avatars), and alert patterns (blinking or color changes). The context space incorporates factors such as the environment (gamified or professional settings), user groups (novices or experienced users), transition types (portals or links), and transition characteristics (frequency and duration). These insights guided the development of five distinct security indicators.

Building on this framework, the authors developed and tested five security indicators: an audio signal using a warning beep for insecure portals, a haptic indicator mimicking heartbeats, a blinking visual indicator in the user's periphery, a static peripheral visual indicator, and a static visual indicator placed above the portal. Each indicator was designed to achieve a balance between clarity, effectiveness, and user acceptability in VR settings. A study involving 25 participants evaluated these indicators within a VR maze game simulating transitions between secure and insecure portals. Participants assessed the indicators on various metrics, including task completion time, accuracy, usability, intrusiveness, disturbance, noticeability, understandability, and personal preferences. This evaluation provided valuable insights into the strengths and weaknesses of each security indicator.

The article highlights the effectiveness of visual indicators, particularly static designs, for their clarity, usability, and minimal intrusiveness, making them the preferred choice for VR security mechanisms. While blinking indicators excelled in accuracy and task completion, they were perceived as distracting, and haptic indicators showed potential as secondary feedback mechanisms but required refinement and improved user familiarity. Audio signals were generally disliked for their intrusiveness, suggesting their utility as complementary rather than primary cues in multi-modal designs. The findings underscore the need for contextually appropriate, well-placed indicators that balance user attention between navigation and security evaluation. The article provides a framework for intuitive and non-intrusive security mechanisms and recommends future research into user adaptation, multi-modal combinations, and optimizing designs for dynamic VR environments. Future work could explore the performance of such indicators in more naturalistic settings while also examining the effects of task type and indicator placement on resumption lag.

8.3.3 Scheme Category 3: Cryptographic and Decentralized Security Solutions and Trust Mechanisms in the Metaverse

The decentralized architecture of the Metaverse necessitates advanced cryptographic solutions to ensure secure, interoperable interactions. This category highlights approaches such as blockchain-based identity management systems, cryptographically signed SWOs for decentralized communication, and reputation-based frameworks to prioritize trustworthy IoT devices in vehicular Metaverse scenarios. These mechanisms collectively address the challenges of trust, integrity, and decentralization, fortifying the Metaverse's infrastructure against emerging threats.

As adversarial VR designs expose significant privacy vulnerabilities by exploiting telemetry data and interactive elements, the broader Metaverse faces challenges safeguarding systemic integrity and user trust. While adversarial threats highlight the need for privacy-preserving mechanisms within immersive environments, Article [7] discusses cryptographic solutions tailored for Metaverse environments as a necessary tool for mitigating privacy risks. These mechanisms could safeguard user data while ensuring the integrity and authenticity of interactions within the Metaverse. For example, decentralized identity management systems powered by blockchain could provide secure, verifiable, and privacy-preserving identity verification processes. Furthermore, enforcing strict access controls and employing end-to-end encryption for communications within Metaverse ecosystems can prevent unauthorized surveillance and data breaches. The necessity of a secure Metaverse infrastructure extends beyond individual user protection to include broader systemic stability according to S. Rostami [7]. Virtual economies within the Metaverse rely heavily on blockchain and Distributed Ledger Technologies (DLTs) for transactions involving Non-Fungible Tokens (NFTs) and digital assets. Security breaches in these financial systems could undermine user trust and disrupt the virtual economy.

While cryptographic solutions offer essential protections for user data and virtual economies, they are only part of the solution to the Metaverse's broader structural challenges. Article [5] addresses these issues by focusing on the foundational architecture of the Metaverse, which builds on the foundational principles of the web as a hypermedia system, where data objects are identified and accessed through Uniform Resource Identifiers (URIs) [5]. However, current Metaverse implementations, which rely heavily on centralized web technologies, face significant scalability, efficiency, and interoperability challenges. These systems often view the Metaverse as an overlay application dependent on centralized infrastructure, which reinforces data silos and creates inefficiencies in cross-platform interactions [5].

Article [5] identifies a semantic gap between the data-oriented application layer and the connection-oriented transport layer in current web implementations. This gap is particularly evident in Metaverse applications requiring real-time data sharing, such as collaborative 3D environments, where centralized client-server communication models limit object-level interoperability and increase latency in scenarios such as shared AR experiences. To address these limitations, the authors propose reimagining the Metaverse as a decentralized, data-oriented hypermedia system that emphasizes secure object-level interoperability. They advocate for a data-centric networking approach aligned with the inherently data-oriented nature of Metaverse applications and introduce SWOs as a foundational building block for this paradigm. SWOs are cryptographically signed data objects with unique identifiers that ensure integrity, authenticity, and ownership. Unlike traditional data packets relying on transport protocols such as HTTP, SWOs operate independently of the communication channel, enabling secure interactions across diverse applications and platforms without requiring back-end system standardization. Each SWO maintains cryptographic integrity, preventing unauthorized modifications or tampering, and supports peer-to-peer interactions and in-network caching, which reduce latency and improve scalability.

A key innovation in this approach is local-first communication, which minimizes dependency on global networks. Users can interact within localized environments using SWOs, enabling secure, real-time collaboration even in offline or disconnected scenarios. This method mitigates latency and enhances system resilience by prioritizing local data synchronization. The potential of this approach is demonstrated through functional prototypes, including the Microverse Editor and the NDN Workspace. The Microverse Editor allows users to collaboratively edit 3D scenes in real time, leveraging SWOs to represent objects, updates, and changes. This prototype highlights features such as cryptographic signing to ensure authenticity, object-level updates to optimize bandwidth, and cross-platform interaction enabled by standardized naming conventions and cryptographic protocols. These features underscore the ability of SWOs to enable secure, efficient, and scalable collaboration in complex 3D environments.

The NDN Workspace, another prototype, applies SWOs in collaborative document editing using Named Data Networking (NDN) principles to create a decentralized, Conflict-free Replicated Data Type (CRDT) system. This workspace supports offline and online functionality, enabling users to edit documents locally without continuous network connectivity while ensuring changes synchronize when the network is restored. The system guarantees eventual consistency, with SWOs ensuring all replicas converge to a consistent state even in distributed environments. Additionally, NDN principles enable in-network caching for efficient data retrieval, reducing latency and bandwidth demands. These prototypes highlight the potential of SWOs in facilitating decentralized, secure, and efficient data sharing across diverse platforms. Figure 8.3 illustrates a collaborative editing scenario among three different users.

Despite these advancements, article [5] identifies several challenges that must be addressed. The complexity of naming and organizing SWOs across different applications presents barriers to universal adoption. Performance issues arise in latency-sensitive applications where cryptographic operations and in-network caching may impact response times. Furthermore, ensuring user-friendly implementations that integrate seamlessly with existing workflows remains critical. The lack of standardized protocols for SWO integration across platforms poses interoperability challenges, complicating widespread adoption.

To address these challenges, the authors propose a research agenda for developing a general SWO web platform based on ongoing work in Named Data Networking. This agenda focuses on creating a robust naming system for SWOs to ensure scalability and efficient object management in the Metaverse. The proposed naming system must support hierarchical, human-readable names that balance simplicity with scalability, ensure global uniqueness to prevent collisions, and enable seamless discovery and retrieval of objects across platforms. Additionally, safeguarding SWOs to ensure trust and integrity in the Metaverse is emphasized. Cryptographic techniques are employed to protect SWOs against tampering and unauthorized access, while fine-grained access control mechanisms provide dynamic permissions. Decentralized Identifiers (DIDs) are highlighted as essential for secure identity management without reliance on centralized systems,

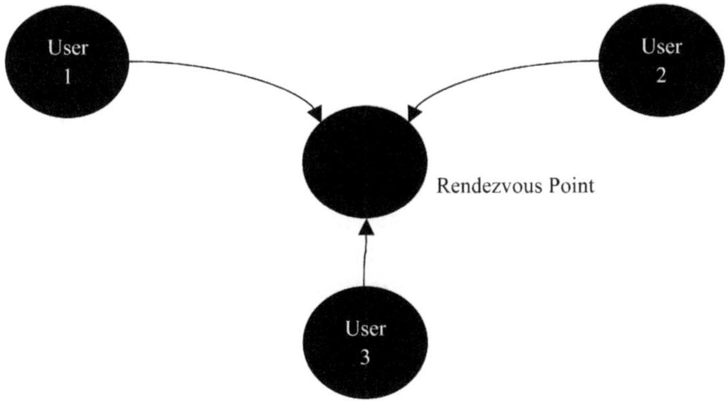

Figure 8.3 Illustration of a collaborative editing scenario among users.

and privacy-preserving methods such as homomorphic encryption are explored to maintain confidentiality during data processing.

The vehicular Metaverse introduces additional layers of complexity, integrating technologies such as VR, AR, AI, and XR within dynamic, data-driven environments. The reliance on real-time data from Sensing IoT (SIoT) devices for creating digital twins adds unique challenges, particularly concerning data integrity and system trustworthiness [4]. As seen in VR security research, effective mechanisms must address technological performance and potential vulnerabilities, such as the moral hazards that arise post-agreement in Sensing Internet of Things (SIoT) participation. Despite advancements in incentivizing SIoT participation using auction and contract theories to ensure Individual Rationality (IR) and Incentive Compatibility (IC), challenges like moral hazards persist. Unlike adverse selection, which occurs before contract formation, moral hazards arise after agreements when malicious devices transmit falsified or outdated data [4]. This behavior can severely degrade the Metaverse's Quality of Service (QoS) and Quality of Experience (QoE) and even lead to life-threatening real-world consequences. Existing Metaverse security measures, such as Moving Target Defenses (MTD), primarily address issues such as Distributed Denial of Service (DDoS) attacks but fail to mitigate poisoning attacks resulting from moral hazards. To address this gap, the paper introduces a novel reputation-based mechanism that employs subjective logic to manage SIoT behavior. This approach integrates vanishing reputation systems to maintain an updated trust environment and mitigate poisoning attacks while retaining reliable devices.

The vehicular Metaverse ecosystem is conceptualized as a marketplace coordinated by a virtual service provider (VSP) that facilitates interactions between SIoT devices and vehicular Metaverse users (VMUs) [4]. The system involves three primary entities: the VSP, a set of SIoT devices that provide data, and VMUs that consume digital twin services derived from SIoT data. Each SIoT device is equipped with sensors for geospatial data collection and machine learning models to extract semantic information transmitted to the VSP. Semantic data transmission optimizes network efficiency by reducing redundancy. A reverse auction mechanism allows SIoT devices to submit bids detailing their data's semantic value and costs, enabling the VSP to select trustworthy devices for constructing the digital twin. While this mechanism effectively addresses adverse selection by filtering unsuitable data, it does not mitigate moral hazards, where devices may intentionally transmit malicious or outdated data post-selection.

The system addresses three major types of poisoning attacks. The first is the service cost reduction attack, where malicious SIoT devices transmit outdated semantic data, conserving energy while compromising the digital twin's quality. The second is the digital twin tampering attack, involving intentional modifications to semantic data, such as altering object types or locations, which can disrupt services or cause physical accidents. The third is the man-in-the-middle attack, where adversaries exploit SIoT vulnerabilities to replace legitimate data with manipulated inputs, often bypassing detection by the VSP. To counter these threats, the proposed model uses subjective logic to evaluate SIoT device trustworthiness. Feedback from VMUs dynamically updates reputation scores, ensuring reliable devices are prioritized while malicious participants are excluded. This feedback loop enhances the system's robustness against poisoning attacks. An illustration of the concept is given in Figure 8.4.

The reputation-based framework calculates SIoT device trustworthiness using subjective logic, which incorporates three components: belief, disbelief, and uncertainty. VMUs provide feedback on digital twin quality, weighted by a fading parameter to prioritize recent interactions. reputation value of SIoT device i from VMU j is calculated as, $R_{i:j}^{fin} = b_{i:j}^{fin} + \alpha_i u_{i:j}^{fin}$ where $b_{i:j}, d_{i:j}, u_{i:j}, \alpha_i$ represent belief, disbelief, uncertainty, and the effective degree of uncertainty effect on the reputation of SIoT device i, respectively. $b_{i:j} + d_{i:j} + u_{i:j} = 1$ and $b_{i:j}, u_{i:j}, \alpha_i \epsilon [0,1]$. The final reputation value of a SIoT device i would be calculated as the average value of the reputation values received from all (M number of VMUs) VMUs, i.e.,

$$R_i^{fin} = \frac{\sum_{j=1}^{M} R_{i:j}^{fin}}{M} \tag{8.3}$$

Reputation scores evolve iteratively as feedback values are aggregated and adjusted for temporal relevance, ensuring that recent behavior has a greater impact on a device's reputation. A reputation backpropagation mechanism maps VMU feedback to specific regions within a three-dimensional SIoT coverage grid to mitigate inaccuracies from sparse or misleading feedback. Negative feedback is attributed only to devices associated with degraded performance in those regions, while semantic consistency checks among co-located devices refine the system's assessments. Outliers with low semantic similarity to their peers are penalized, while other devices receive positive reinforcement. The reverse auction mechanism further enhances system efficiency by maximizing social welfare – the aggregate utility derived from interactions between the VSP and SIoT devices. Utility for the VSP is determined by the value of collected semantic data minus associated costs,

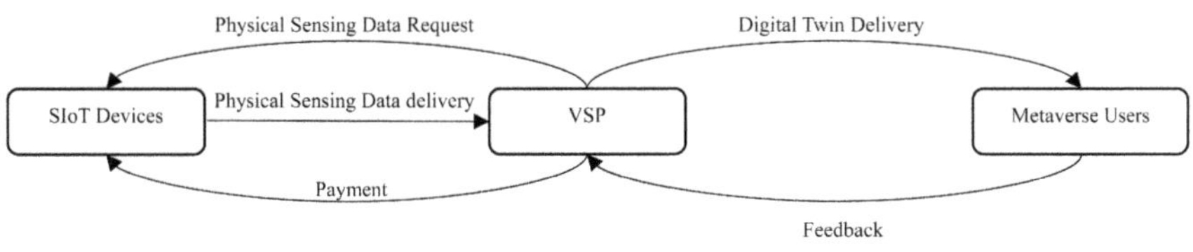

Figure 8.4 Proposed system model in [4].

while SIoT devices calculate utility as payments received minus operational expenses. The auction mechanism balances these utilities while integrating reputation scores to prioritize trustworthy devices, ensuring robust performance even under adversarial conditions. This mechanism is formulated as an integer linear programming problem to optimize resource allocation and prevent manipulation by malicious actors.

The authors performed extensive experiments to validate the proposed reputation-based mechanism using real-world data from the CARRADA dataset, which includes synchronized camera and radar imagery annotated for semantic segmentation. Simulations model a Metaverse market with 20 SIoT devices and a fixed channel capacity of 10 kbps per device. The dataset captures diverse environmental conditions and device perspectives, allowing for a realistic evaluation. Over 10 simulation episodes, each comprising 50 time steps, three device types with varying behavioral patterns are tested: Type-1 devices (honest), Type-2 devices (frequently malicious), and Type-3 devices (alternating between benign and malicious behavior).

The results demonstrate the mechanism's effectiveness in mitigating poisoning attacks. Without the mechanism, indiscriminate SIoT selection results in high attack rates and degraded digital twins. By prioritizing trustworthy devices, the reputation mechanism significantly reduces tampering incidents. Type-3 devices, which exhibit intermittent malicious behavior, are included more often than consistently malicious Type-2 devices, showcasing the system's nuanced evaluation. Feedback filtering enhances accuracy by preventing reliable devices from being excluded due to false negatives, though this conservative selection approach slightly reduces overall social welfare. The trade-off reflects the system's prioritization of security over raw efficiency. Overall, the mechanism achieves a balance between trustworthiness, efficiency, and security in the vehicular Metaverse. Future work should explore identifying the optimal weights to reduce the influence of previous data.

8.4 COMPARISON AND FUTURE TRENDS

To comprehensively compare the three described schemes, the following Table 8.1 organizes their key attributes across critical dimensions. This comparison highlights each scheme's focus area, the main challenges it seeks to address, the key approaches employed, the findings derived from experimental or theoretical studies, and the broader implications of their implementation in enhancing the security, privacy, and trustworthiness of Metaverse environments. The table serves as a consolidated view of these strategies, facilitating a clear understanding of their relative strengths and contributions.

Table 8.1 Comparison of security and privacy schemes in the Metaverse across key dimensions

Key Features	*Scheme Category 1*	*Scheme Category 2*	*Scheme Category 3*
Focus Area	Privacy and vulnerability in VR systems.	Machine learning models for security and user-friendly safeguards in VR.	Cryptographic methods and decentralized frameworks to secure Metaverse interactions.
Main Challenges	Adversarial designs exploiting telemetry data to infer sensitive user attributes such as demographics, cognitive traits, and anthropometrics.	Balancing interpretability with high detection accuracy; mitigating data imbalance in attack classifications; addressing firmware-specific vulnerabilities and usability concerns for security indicators.	Addressing scalability, interoperability, and latency issues; mitigating moral hazards in IoT-driven systems; ensuring user-friendly integration and consistent object-level security across diverse platforms.
Key Approaches	Escape room-style VR environments; telemetry data collection via motion, audio, and spatial inputs using devices such as HTC Vive and Oculus Quest 2.	DNN-based Network Intrusion Detection Systems (NIDS), BiLSTM frameworks for vulnerability detection, SHAP/LIME for model interpretability.	Blockchain for decentralized identity, Secure Web Objects (SWOs) for interoperable data sharing, and reputation-based mechanisms for IoT trust.
Findings	High accuracy in extracting user attributes (e.g., height, handedness, demographics, etc.) using telemetry data and machine learning models.	High detection accuracy for attacks such as DDoS and SQL Injection using DNNs; VRVul-Discovery BiLSTM identified 7 undocumented vulnerabilities in VR firmware.	SWOs enable secure, scalable real-time collaboration and offline synchronization. Reputation-based mechanisms effectively mitigate poisoning attacks in IoT systems, enhancing trust and system reliability.
Implication	Raises concerns about surveillance, manipulation, and compliance with data privacy laws. Recommends ethical guidelines and advanced countermeasures such as differential privacy.	Combines technical robustness with user-friendly features to enhance trust, usability, and proactive threat mitigation.	Enhances system integrity, scalability, and cross-platform interoperability while addressing challenges such as naming complexity, performance, and standardization.

The future of privacy in VR and the security of the Metaverse will be defined by the convergence of advanced technologies, ethical practices, and resilient frameworks. Privacy-preserving methods, such as differential privacy and obfuscation, are set to safeguard sensitive data by introducing controlled noise, protecting user identities without undermining system performance. Privacy-aware APIs and SDKs will play a critical role in promoting ethical design and minimizing unnecessary data exposure, while regulatory advancements will emphasize real-time consent mechanisms, transparent data usage policies, and ethical AI auditing to ensure fairness and compliance. Innovations in local data processing and privacy-focused tracking systems will further mitigate risks associated with biometric data transmission, empowering users with tools and knowledge to manage their personal data, particularly as VR expands into sensitive domains such as healthcare and education. Simultaneously, Metaverse security advancements will prioritize AI-driven solutions for robust identity verification and anomaly detection to combat threats such as avatar de-anonymization and malicious payloads. Explainable AI (XAI) will enhance the transparency of network intrusion detection systems, while user-centric security features, such as visual and haptic cues, will bolster trust and decision-making in virtual environments. Decentralized architectures, including cryptographic solutions such as SWOs and blockchain technologies, will ensure interoperability and data integrity, supporting token-based economies and secure transactions. Defense strategies such as Moving Target Defense (MTD), predictive AI models, and 6G-enabled ultra-reliable low-latency communication (URLLC) will address emerging threats such as adversarial AI and DDoS attacks. Together, these innovations will establish scalable, adaptive, and transparent privacy and security frameworks, supporting the Metaverse's expansion in education, commerce, and social interaction while reinforcing user trust.

8.5 CONCLUSION

The chapter delves into the intricate and multifaceted challenges of securing the Metaverse, an evolving digital ecosystem that seamlessly integrates physical and virtual realities. By categorizing security strategies into privacy protection, machine learning-driven safeguards, and cryptographic and decentralized solutions, the chapter highlights the innovative approaches required to mitigate risks inherent in immersive environments. Privacy vulnerabilities, such as adversarial designs leveraging telemetry data, underscore the necessity for robust countermeasures, including differential privacy and ethical guidelines. Advanced machine learning frameworks and intuitive security indicators exemplify the critical balance between technical rigor and user-centric design in securing VR systems. Meanwhile, blockchain-based decentralized mechanisms and cryptographically SWOs promise scalable and interoperable solutions to systemic challenges such as trust and integrity in Metaverse interactions. Looking forward, the convergence of ethical practices, privacy-preserving technologies, AI-driven detection systems, and decentralized architectures will shape the Metaverse's security landscape. By fostering user trust and safeguarding sensitive data, these advancements will support the Metaverse's transformative potential across domains such as education, commerce, and social interaction.

REFERENCES

1. V. Nair, G. M. Garrido, D. Song, and J. O'Brien, "Exploring the Privacy Risks of Adversarial VR Game Design," *Proc. Priv. Enhancing Technol.*, 2023, Accessed: Nov. 27, 2024. [Online]. Available: https://petsymposium.org/popets/2023/popets-2023-0108.php
2. L. Sha *et al.*, "VRVul-Discovery: BiLSTM-based Vulnerability Discovery for Virtual Reality Devices in Metaverse," *ACM Trans Multimed. Comput Commun Appl*, Jul. 2024. doi: 10.1145/3677609.
3. E. C. Nkoro, C. I. Nwakanma, J.-M. Lee, and D.-S. Kim, "Detecting Cyberthreats in Metaverse Learning Platforms Using an Explainable DNN," *Internet Things*, vol. 25, p. 101046, Apr. 2024. doi: 10.1016/j.iot.2023.101046.
4. I. Lotfi, M. Qaraqe, A. Ghrayeb, and N. Dusit, "Enhancing Trust and Security in the Vehicular Metaverse: A Reputation-Based Mechanism for Participants with Moral Hazard," May 23, 2024, *arXiv*: arXiv:2405.19355. doi: 10.48550/arXiv.2405.19355.
5. T. Yu *et al.*, "Secure Web Objects: Building Blocks for Metaverse Interoperability and Decentralization," Jul. 21, 2024, *arXiv*: arXiv:2407.15221. doi: 10.48550/arXiv.2407.15221.
6. M. Windl, A. Scheidle, C. George, and S. Mayer, "Investigating Security Indicators for Hyperlinking within the Metaverse," presented at the Nineteenth Symposium on Usable Privacy and Security (SOUPS 2023), 2023. Accessed: Nov. 27, 2024. [Online]. Available: https://www.usenix.org/conference/soups2023/presentation/windl
7. S. Rostami, "Experimental Investigation of Advanced Metaverse Realms of Experience.," Phd, Doctorat en télécommunications, Québec, 2022. Accessed: Nov. 30, 2024. [Online]. Available: https://espace.inrs.ca/id/eprint/13244/

Chapter 9

Attacks in Metaverse Platforms

Recent Trends

Leo Yang and Kenny Ricks

9.1 INTRODUCTION

The Metaverse is redefining digital interaction, enabling users to immerse themselves in virtual environments and conduct activities ranging from social interactions to economic transactions. With advancements in technologies such as extended reality (XR), artificial intelligence (AI), and blockchain, the Metaverse is becoming increasingly complex and integrated into everyday life. However, this technological sophistication introduces unprecedented security challenges, including identity theft, cyberattacks, and data breaches.

Securing the Metaverse is essential not only to protect user data and digital assets but also to foster trust and adoption among users. This chapter aims to:

1. Discuss the motivations for enhancing Metaverse security
2. Examine three popular attack models on Metaverse hardware and software
3. Present contemporary countermeasures categorized into prevention, detection, and mitigation strategies
4. Highlight future research trends and challenges in Metaverse security

9.2 MOTIVATION FOR SECURITY IN THE METAVERSE

The Metaverse not only brings forth a range of transformative applications but also exposes users to unprecedented security risks. The motivations for enhancing its security are rooted in the need to protect identity, privacy, assets, and user safety while ensuring trust in the platform.

9.2.1 Identity Verification and Traceability

Avatars, which represent users in the Metaverse, are vulnerable to impersonation, identity theft, and manipulation. To ensure trust, robust frameworks

 DOI: 10.1201/9781032663609-11

must be developed. For example, two-factor authentication models combining biometric-based systems and chameleon signatures offer enhanced traceability, linking virtual identities with physical counterparts securely [1]. Blockchain-based identity management systems also provide decentralized solutions, enabling immutable and verifiable identity records [2].

9.2.2 Privacy Preservation

Immersive experiences in the Metaverse rely on collecting granular user data, including biometrics, facial expressions, and behavioral patterns. While this enhances realism, it also exposes users to privacy breaches. Misuse of sensitive data can result in identity theft, surveillance, and unauthorized profiling. Ensuring data protection through decentralized storage, encryption, and anonymization techniques is vital to preserving user trust and compliance with privacy regulations [3, 4].

9.2.3 Protection Against Cyber Threats

The integration of heterogeneous technologies within the Metaverse introduces vulnerabilities to advanced cyber threats, including Denial of Service (DoS) attacks – disrupting virtual environments and user interactions, advanced persistent threats (APTs) – exploiting unpatched vulnerabilities for long-term exploitation, And AI-driven attacks – using adversarial AI to bypass traditional defenses [3, 4]. AI-driven intrusion detection systems (IDS) and automated threat mitigation tools are imperative for proactive defense mechanisms in the Metaverse.

9.2.4 Economic Security

The economic infrastructure of the Metaverse, underpinned by blockchain, cryptocurrencies, and NFTs, is an attractive target for cybercriminals. Incidents such as the theft of high-value NFTs highlight the urgency for secure asset management and transaction systems. Blockchain-integrated security measures and AI-based fraud detection are critical to safeguarding the economic integrity of the Metaverse [2, 4].

9.2.5 Decentralized Architecture and Interoperability

The decentralized nature of the Metaverse promotes autonomy but complicates trust management and accountability. Ensuring consensus among diverse entities without a central authority requires robust cryptographic frameworks. Blockchain technologies and interoperable standards are essential to achieving seamless and secure interactions across different Metaverse platforms [2].

9.2.6 User Safety and Behavioral Security

The Metaverse mirrors real-world interactions, including the potential for malicious behaviors such as harassment and stalking. Reports of virtual crimes have raised concerns about user safety. Ethical AI frameworks and real-time monitoring systems can help identify and mitigate inappropriate behaviors, fostering a secure and inclusive environment [1, 5].

9.2.7 Trustworthy AI Systems

AI is integral to the Metaverse, powering applications such as NPCs, digital twins, and personalized experiences. However, untrustworthy AI can exploit vulnerabilities, leading to biased outcomes or malicious manipulation. Ensuring transparency, accountability, and robustness in AI systems is essential for sustainable development [2, 4].

9.3 POPULAR ATTACK MODELS ON THE METAVERSE HARDWARE AND SOFTWARE UNITS

9.3.1 Man-in-the-Middle Attacks

Man-in-the-middle (MITM) attacks involve intercepting and potentially altering communication between two entities, such as users and servers. These attacks exploit vulnerabilities in communication protocols, targeting sensitive information such as user credentials, transaction details, or private messages.

(1) Main Features:

Interception: The attacker captures the data transmitted between endpoints.

Modification: The intercepted data can be altered to deceive users or inject malicious commands.

Deception: Users are unaware of the attack, enabling the stealthy exploitation of their data.

(2) Mathematical Model

Let S represent the sender, R the receiver, and A the attacker. The legitimate communication $C(S,R)$ is replaced by $C'(S,A,R)$, where:

$$C'(S,A,R) = \begin{cases} f(C(S)), \textit{if A modeifies the data,} \\ C(S), \textit{if no modification occurs} \end{cases}$$

Here, $f(C(S))$ represents the altered communication by the attacker.

Figure 9.1 demonstrates the attacker intercepting the data flow between a user and a server, showing how the data can be manipulated to cause harm.

(3) Countermeasures for MITM Attacks

Secure Protocols: Employing end-to-end encryption and secure communication protocols such as TLS.

Authentication Mechanisms: Multi-factor authentication (MFA) to ensure legitimate connections.

Network Monitoring: Real-time monitoring to detect anomalies in data transmission.

9.3.2 Adversarial Machine Learning Attacks

Adversarial attacks target AI models used in Metaverse platforms for tasks such as object detection, avatar generation, and decision-making. By introducing minor perturbations to input data, attackers can mislead AI models into making incorrect classifications or decisions.

(1) Main Features

Perturbation: Small, often imperceptible modifications to input data cause significant errors.

Targeted vs. Untargeted: Targeted attacks produce specific incorrect outputs, while untargeted attacks aim to generate any misclassification.

Exploitation: Vulnerabilities in AI models are exploited to compromise system functionality.

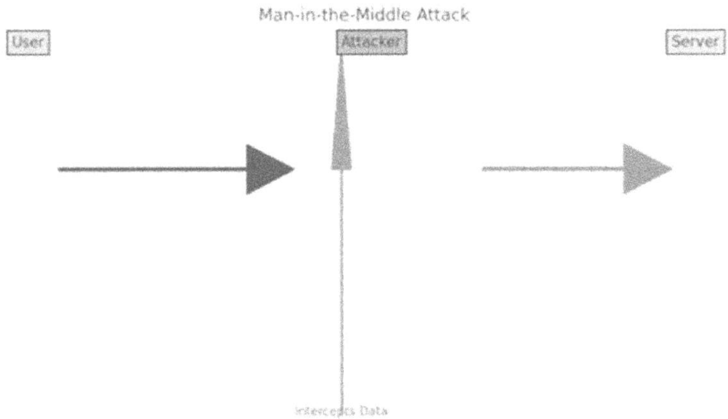

Figure 9.1 MITM attacks.

(2) Mathematical Model

Given an AI model $f(x)$, and an adversarial example x' is crafted as: $x' = x + \delta$,

where δ is a perturbation designed to maximize the model's error:

$$\text{maximize } f(x) - f(x')$$

subject to $' < \epsilon$, ensuring perturbations remain small.

Figure 9.2 highlights an original input, a perturbed version, and the resulting model misclassification, emphasizing the model's vulnerability to adversarial inputs.

(3) Countermeasures for Adversarial ML Attacks

Robust Training: Incorporating adversarial training to improve the resilience of AI models.

Detection Systems: Deploying systems to identify and mitigate adversarial inputs.

Model Validation: Regular testing against known adversarial patterns.

9.3.3 Ransomware Attacks on Virtual Assets

Ransomware attacks involve encrypting user data, including virtual assets such as avatars, NFTs, and virtual properties, and demanding payment for decryption. This attack targets both hardware storage devices and software-based asset management systems.

(1) Main Features

Encryption: The attacker locks user data with strong cryptographic algorithms.

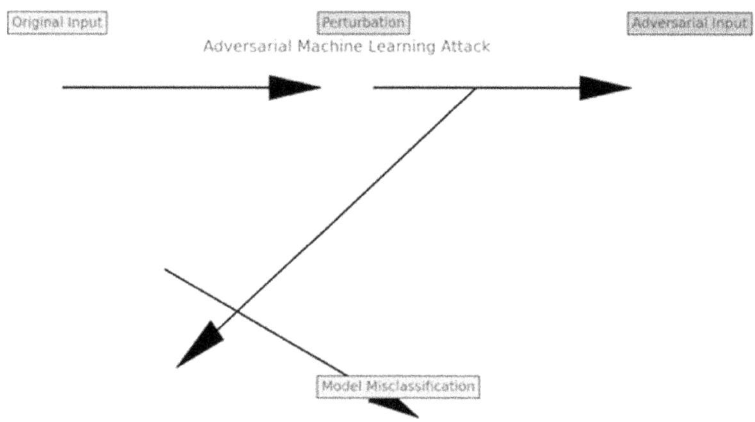

Figure 9.2 Adversarial machine learning attacks.

Ransom Demand: Victims must pay a ransom, typically in cryptocurrency, to retrieve their data.
Data Loss: Failure to comply with the ransom demand may lead to permanent loss of data.

(2) Mathematical Model
Let D represent the user's data, and K be the encryption key. The attacker encrypts the data as:
$E(D,K) = D_{encrypted}$.
The decryption key K' is withheld until a ransom R is paid:

$$\text{Decrypt}\left(D_{\text{encrypted}}, K'\right) \text{iff R paid}$$

Figure 9.3 illustrates how user data is encrypted and held hostage, with a ransom note demanding payment for decryption.
(3) Countermeasures for Ransomware Attacks
Regular Backups: Frequent data backups to mitigate the impact of ransomware.
Access Control: Restricting unauthorized access to critical systems and data.
Awareness Programs: Educating users about phishing and social engineering tactics.

9.4 COUNTERMEASURES AGAINST METAVERSE ATTACKS

9.4.1 Prevention of Attacks

9.4.1.1 Zero-Trust Architecture (ZTA)

Role: Enforces strict verification for every access request, regardless of origin, minimizing trust assumptions within the network.
Solution: Implementing ZTA ensures that all entities are authenticated and authorized before accessing resources, reducing the risk of unauthorized access [6]. Please refer to Figure 9.4 for its workflow,
Mathematical Model: Access is granted if and only if:

$$\text{Access} = \begin{cases} \textit{Allow} \quad \textit{if Auth}(\text{u}) \wedge \textit{Policy}(u,r) \\ \textit{Deny} \qquad\qquad\qquad\qquad \textit{otherwise} \end{cases}$$

where $\text{Auth}(\text{u})$ verifies user u's identity, and $\text{Policy}(\text{u},\text{r})$ checks if user u has rights to resource r.

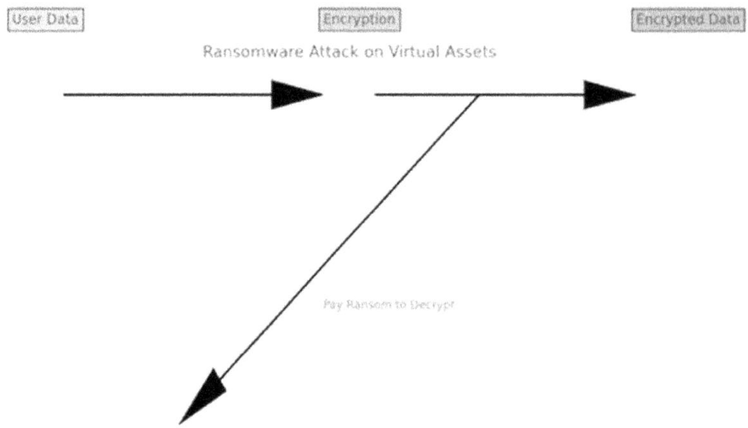

Figure 9.3 Ransomware attacks on virtual assets.

9.4.1.2 Blockchain for Data Integrity

Role: Provides a decentralized ledger to ensure the immutability and transparency of transactions and data exchanges.

Solution: Utilizing blockchain technology secures digital assets and transaction records, preventing tampering and fraud [7]. Its workflow is shown in Figure 9.5.

Mathematical Model: Each block B_i contains:

$$B_i = \{Data_i, H(B_{i-1}), Nonce_i\}$$

where $H(B_{i-1})$ is the hash of the previous block, ensuring a tamper-evident chain.

9.4.2 Detection of Attacks

9.4.2.1 AI-Based Intrusion Detection Systems (AI-IDS)

Role: Leverages machine learning algorithms to identify anomalous patterns indicative of cyber threats.

Solution: AI-IDS monitors network traffic and user behavior, enabling real-time detection of suspicious activities [4], as shown in Figure 9.6.

Mathematical Model: Anomaly score S is computed as:

$$S = f(x) - \mu$$

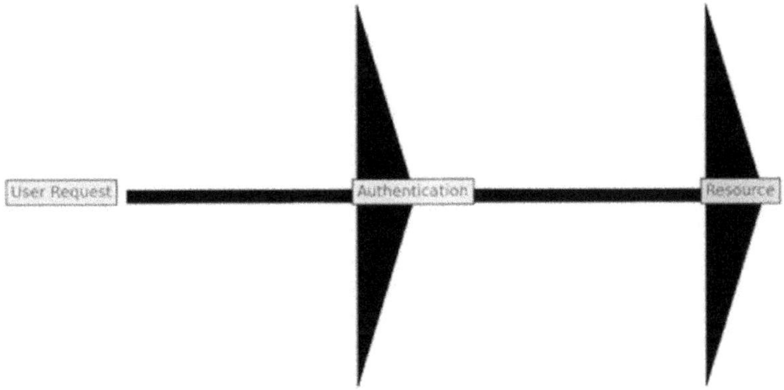

Figure 9.4 ZTA workflow.

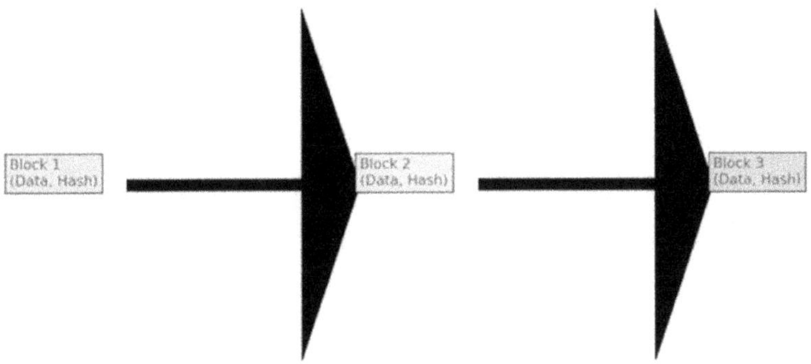

Figure 9.5 Blockchain data structure.

where $f(x)$ is the observed behavior, and μ is the expected behavior. An alert is triggered if $|S| > \mathit{threshold}$.

9.4.2.2 Multimodal Biometric Authentication

Role: Combines multiple biometric factors (e.g., facial recognition, voice patterns) for continuous user verification.

Solution: Enhances security by ensuring that access is granted only to authenticated users, reducing the risk of impersonation [8], as shown in Figure 9.7.

Mathematical Model: Authentication score A is calculated as:

$$A = \omega_1 \cdot FR + \omega_2 \cdot VP$$

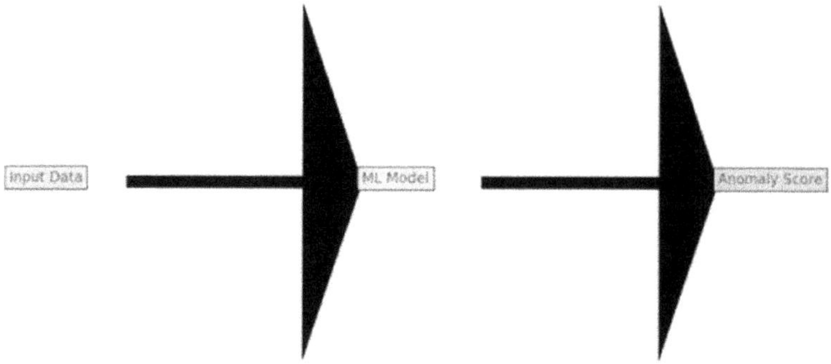

Figure 9.6 AI-based intrusion detection system.

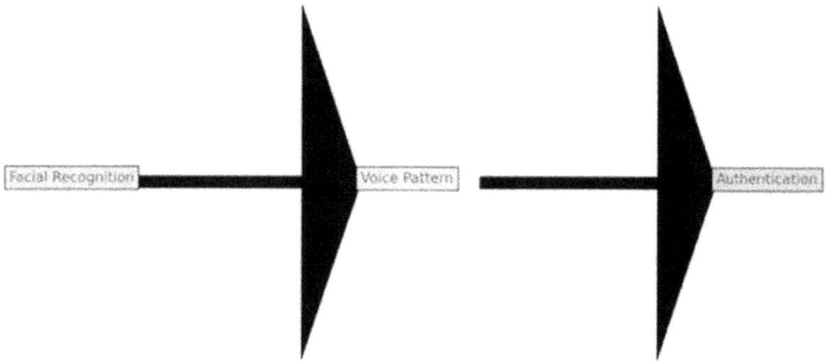

Figure 9.7 Multimodal biometric authentication process.

where FR is the facial recognition score, VP is the voice pattern score, and ω_1, ω_2 are weights summing to 1. Access is granted if $A \geq threshold$.

9.4.3 Mitigation of Attacks

9.4.3.1 *Self-Healing Networks*

Role: Automatically detect and respond to network anomalies, restoring normal operations without human intervention.

Solution: Employs AI to identify and isolate compromised components, ensuring resilience against attacks [9], as shown in Figure 9.8.

Mathematical Model: Network state N transitions from compromised C to healthy H based on healing function h:

$$N(t+1) = h(N(t))$$

where h applies corrective actions to restore N to H.

9.4.3.2 *Decentralized Storage and Backup*

Role: Distributes data across multiple nodes to prevent single points of failure and facilitate data recovery.
Solution: Utilizes decentralized storage solutions to ensure data availability and integrity, even during attacks [7].
Mathematical Model: Data D is split into n fragments D_i and distributed across nodes. Reconstruction is possible if a quorum q of fragments is available:

$$D = \sum_{i=1}^{q} D_i \quad .$$

Ensuring data recovery despite node failures.

9.5 FUTURE RESEARCH TRENDS AND CHALLENGES IN METAVERSE SECURITY

9.5.1 AI-Powered Adaptive Security

AI is pivotal for enabling real-time and adaptive security in the Metaverse. Future research will focus on building intelligent systems capable of detecting, predicting, and mitigating sophisticated threats.

AI models leveraging deep learning and anomaly detection can continuously monitor Metaverse activity for irregular patterns. These systems can differentiate between benign anomalies (e.g., sudden user behavior changes) and malicious activities.

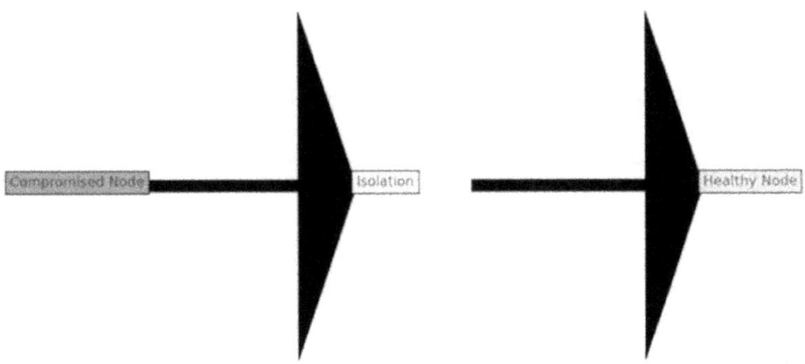

Figure 9.8 Self-healing networks mechanism.

Future research will explore explainable AI (XAI) to make detection outcomes interpretable, which is crucial for user trust and regulatory compliance.

Reinforcement learning algorithms can be trained to dynamically respond to evolving threats by simulating attack scenarios. This approach enables systems to "learn" effective countermeasures autonomously. Example: A Metaverse server under Distributed Denial of Service (DDoS) attacks could reallocate resources or isolate affected nodes autonomously.

Adversarial attacks targeting AI models may attempt to manipulate inputs to bypass detection. Developing robust defenses against adversarial perturbations and ensuring the resilience of AI models remain key areas of focus.

9.5.2 Privacy-Preserving Technologies

The Metaverse collects vast amounts of sensitive user data, including biometrics, behavioral patterns, and virtual interactions. Privacy-preserving solutions aim to protect user data while maintaining functionality.

Using Homomorphic Encryption (HE): It enables computations on encrypted data without decrypting it. For example, virtual environments could process encrypted user movements for rendering without accessing raw data. Future research will focus on improving the computational efficiency of HE to enable real-time performance.

Federated Learning (FL): Allows multiple devices to collaboratively train AI models without sharing raw data, reducing privacy risks. For instance, XR headsets could contribute to model updates while keeping user data local. Enhancing FL with differential privacy techniques to further anonymize model updates will be a critical research area.

Challenges: Balancing privacy with usability, particularly for XR devices with limited computational power. Managing trade-offs between real-time performance and the overhead introduced by privacy-preserving mechanisms.

9.5.3 Blockchain-Based Security Frameworks

Blockchain technology is increasingly being explored for decentralized security in the Metaverse, offering transparency, immutability, and trust without centralized authorities.

Decentralized Identity Management: Blockchain can store user identities in decentralized ledgers, ensuring users have full control over their personal data. For instance, users could use blockchain-based IDs to authenticate across platforms without sharing sensitive information.

Future research will focus on self-sovereign identity systems that integrate seamlessly into Metaverse environments.

Secure Virtual Asset Transactions: Blockchain-based smart contracts can automate and secure transactions involving virtual goods, services, and NFTs. Research is needed to address the scalability of blockchain networks, enabling them to handle the high transaction volumes expected in large-scale Metaverse economies.

Cross-Chain Interoperability: Ensuring seamless interaction between different blockchain networks will be essential for enabling asset portability across Metaverse platforms. Future solutions may explore universal protocols for cross-chain communication and asset migration.

Challenges: The energy efficiency of blockchain systems and their scalability for Metaverse-scale applications remain significant obstacles.

9.5.4 Quantum-Resistant Cryptography

The advent of quantum computing poses a threat to traditional cryptographic algorithms, necessitating quantum-resistant cryptographic solutions.

Post-Quantum Cryptography (PQC): Algorithms such as lattice-based cryptography, hash-based signatures, and code-based cryptography offer resistance to quantum attacks. Future research will focus on optimizing PQC algorithms for lightweight implementation in XR devices.

Quantum Key Distribution (QKD): Provides theoretically unbreakable encryption by leveraging the principles of quantum mechanics. QKD can secure communication channels in Metaverse environments. Integrating QKD into existing Metaverse infrastructure will be an area of ongoing exploration.

Challenges: Transitioning existing Metaverse security systems to quantum-resistant standards without disrupting functionality or introducing excessive computational overhead.

9.5.5 Ethical AI and Governance Models

As AI becomes more pervasive in the Metaverse, ethical considerations will play a crucial role in guiding its development and deployment.

Explainable AI (XAI): Ensures that AI decisions in the Metaverse are transparent and interpretable. For example, users should understand why an avatar was flagged for suspicious activity or denied access to a virtual space. Future research will focus on developing lightweight XAI models suitable for real-time applications.

Governance Frameworks: Establishing rules and policies to regulate user interactions, digital ownership, and platform behavior. For example, governance frameworks could enforce community guidelines to prevent virtual harassment. Decentralized autonomous organizations (DAOs) may be explored as governance models for self-regulating Metaverse ecosystems.

Challenges: Balancing regulation with freedom and creativity in the Metaverse and achieving global consensus on ethical AI standards.

9.5.6 Enhanced Cyber-Physical Security

The integration of XR hardware and IoT devices into the Metaverse expands the attack surface, requiring innovative security solutions.

Secure Hardware Design: Future XR devices must incorporate secure boot mechanisms, tamper-proof hardware, and secure firmware updates to prevent unauthorized access. Hardware-level biometric authentication, such as iris or fingerprint scanning, will further enhance security.

Behavioral-Based Authentication: Continuous authentication using behavioral patterns (e.g., typing speed, gaze direction) will ensure only legitimate users access Metaverse environments. AI-based models can analyze subtle deviations in user behavior to detect anomalies.

Challenges: Ensuring secure and seamless integration of low-power devices, such as XR headsets and wearables, without compromising user experience.

9.6 CONCLUSIONS

The Metaverse offers transformative potential, but its security challenges require innovative solutions. By addressing threats such as MITM attacks, ransomware, and adversarial machine learning, and adopting proactive countermeasures, the Metaverse can evolve into a secure and trusted environment. Future research must focus on scalable, interoperable, and sustainable security solutions to address emerging challenges and foster Metaverse adoption.

REFERENCES

1. Yang, Kedi, et al. "A secure authentication framework to guarantee the traceability of avatars in metaverse." *IEEE Transactions on Information Forensics and Security* 18 (2023): 3817–3832.

2. Ryu, Jongseok, et al. "Design of secure mutual authentication scheme for metaverse environments using blockchain." *IEEE Access* 10 (2022): 98944–98958.
3. Wang, Yuntao, et al. "A survey on metaverse: Fundamentals, security, and privacy." *IEEE Communications Surveys & Tutorials* 25.1 (2022): 319–352.
4. Awadallah, Abeer, et al. "Artificial intelligence-based cybersecurity for the metaverse: Research challenges and opportunities." *IEEE Communications Surveys & Tutorials* 27 (2024): 1008–1052.
5. Qayyum, Adnan, et al. "Secure and trustworthy artificial intelligence-extended reality (AI-XR) for metaverses." *ACM Computing Surveys* 56.7 (2024): 1–38.
6. Gupta, Ankur, et al. "Metaverse security: Issues, challenges and a viable ZTA model." *Electronics* 12.2 (2023): 391.
7. Manzoor, Khadija, Umara Noor, and Zahid Rashid. "NFT-based blockchain-oriented security framework for metaverse applications." *arXiv preprint arXiv:2307.10342* (2023).
8. Sharma, S., J. Singh, A. Gupta, F. Ali, F. Khan, and D. Kwak. "User Safety and Security in the Metaverse: A Critical Review." *IEEE Open Journal of the Communications Society* 5 (2024): 5467–5487.

Chapter 10

Security and Privacy for AR/VR Spaces

Shehan Irteza Pranto

10.1 INTRODUCTION

The Metaverse, a futuristic virtual infrastructure, holds immense potential. It offers a realm where people can engage in various activities, from gaming to work, shopping, and socializing, all within virtual environments that mirror real life. This virtual realm is built upon a sophisticated technical infrastructure, incorporating virtual reality (VR), augmented reality (AR), high-speed internet connections, and the motion of virtual space. The Metaverse is set to revolutionize the way we interact, introducing self-sustaining virtual shared spaces and seamlessly integrating the physical and digital worlds, surpassing the current web and internet revolutions [1].

Generally, the components of the metaverse can be classified into six categories: (1) interactivity – the interaction will no longer be held based on mobile or laptop. Instead, the augmented experience in virtual space through AR, VR, and mixed reality will be the major form of interaction [2]. (2) Digital twin – it refers to the identical virtual space of the real world where the physical entities will be cloned by the sensory data and embedded systems with high precision [3]. (3) Networking – the metaverse requires high-speed data transmission with ultra-reliable security between the virtual and real world as well as the infrastructure of 6G technology [4]. (4) Ubiquitous computing – the versatility of interaction not restricted by time or location [5]. (5) Artificial intelligence – it gives the learning ability and decision-making capability to the other components of the metaverse. Thus, it works as the "brain" of the Metaverse [6]. (6) Blockchain – it provides decentralized solutions with hash-chained block structured data to avoid centralization risks such as low transparency and limited entity dependencies [7].

Along with numerous blessings, the metaverse infrastructure also comes with privacy and security concerns. More or less, all the metaverse applications require user authentication and access control processes before offering the services. However, the attackers can steal the user's identity, which may affect the user's physical and digital life more severely than the conventional information system. Besides, the security of an enormous number of

 DOI: 10.1201/9781032663609-12

data produced by the services of the metaverse is another primary concern of metaverse security. For instance, the data from hardware sensors in the healthcare domain may be tempered by injecting false data or changing the existing data by attackers, leading to the wrong treatment of the patients. Additionally, data leakage during wired or wireless communication may lead to users' privacy leakage. So, there is a clear need for continued studies to develop secured metaverse services that ensure the safety, privacy, and security of users. Otherwise, ultra-reliable services with high-speed communication, which is the backbone of metaverse technologies, cannot be possible. Thus, the users may not benefit from the metaverse's next-generation technologies.

The combination of AR and VR presents the physical world in a virtual domain in three-dimensional space in various applications. There is a lack of research on the security of AR/VR spaces for employment in the metaverse. However, strong and comprehensive investigations are required for the secured operation among AR/VR systems since, in the metaverse, this technology acts as the core foundation for representing the physical world in the virtual domain. In [8], the authors represented the visual perturbation that causes injury to VR users, and they also analyzed the harmful effects of such attacks. In this study, we focused on figuring out the vulnerabilities associated with the AR/VR systems in the metaverse that may compromise user privacy and confidentiality. We also discussed the possible solutions that may be incorporated to resolve such issues.

10.2 METAVERSE SYSTEM ARCHITECTURE

The metaverse is a connected infrastructure of human, physical, and digital worlds where users interact with each other in virtual space using their virtual identity as avatars with a high-speed data transmission system. Figure 10.1 represents the general architecture of the metaverse system.

10.2.1 Human World

The human world consists of social interaction between users, and interactions are developed based on human psychology. In simple words, the human world is where users socialize with each other by using virtual space via AR, VR, and MR, use digital avatars as their identity and utilize human-computer interaction (HCI) for play, work, and various tasks in daily life.

10.2.2 Physical World

The physical infrastructure provides a strong communication and transmission system between humans and the physical world. This metaverse

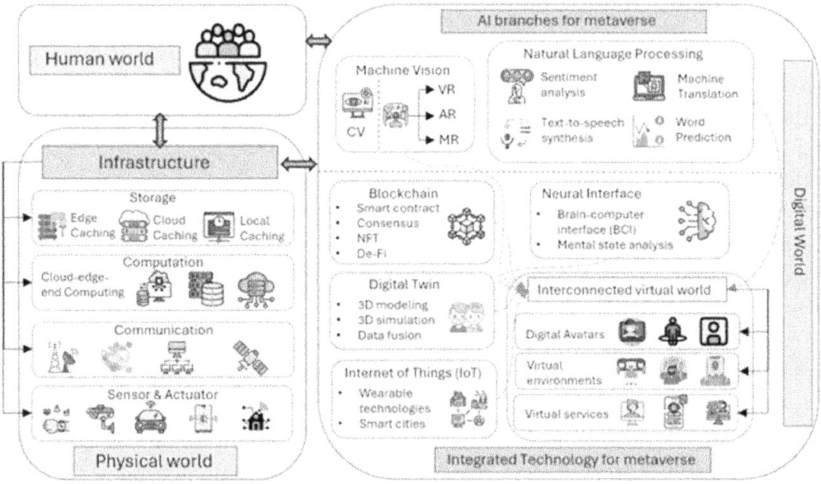

Figure 10.1 Architecture of metaverse system integrated with the human, physical, and digital world.

infrastructure has an efficient, connected network between sensing systems, controlling software, transmission protocol, ubiquitous computation, and storage services. The sensing system developed by various hardware sensors, smart devices, and physical control systems offers data perception from environments such as transportation systems, human bodies, industries, and automotive vehicles. The wired and wireless transmission by satellite, Wi-Fi, and cellular communications are part of the communication subsection of the physical world. Moreover, high-speed computing power is provided by cloud-edge end computing [9]. Finally, the big data storage capacity through edge caching, cloud caching, and local caching enables users to manage enormous amounts of data easily.

10.2.3 Digital World

Digital infrastructure offers the interconnection of various virtual services of the metaverse, including machine vision, natural language processing (NLP), blockchain, neural interface, digital twin, and Internet of Things (IoT); users interact with the digital world by creating a virtual avatar that works as the user's identity. Machine vision and NLP are two AI branches of the metaverse in the digital world. Machine vision includes virtual, augmented, and mixed reality, where the users interact with each other or play games and do various tasks in virtual space using HMDs and BCI devices. The applications of NLP, such as machine translation, remove the language barriers in the metaverse. Besides, customer reviews in e-commerce businesses or other services can be analyzed quickly, and actions can be taken accordingly using the sentiment analysis process. Similarly, text-to-speech synthesis can be used in robot-based domains. Monitoring neural activity by BCI devices helps to understand the user's state of mind for a specific task. For instance, the students' consciousness level in the classroom can be identified by this technology, which may help students increase concentration during class lectures. Another component of the digital world is the IoT, where sensor-based embedded systems, wearable technologies, and data transmission protocols via W-fi or Bluetooth play a vital role in developing smart applications. Digital twins and blockchain are also the components of an interconnected virtual world where the digital twin.

10.3 AI-BASED SECURITY FOR AR/VR IN THE METAVERSE

AR has extended its adaptability in various sectors such as education, healthcare, industry, and defense [10]. Though this technology has blessed us with establishing the connection between the virtual world and the human world, it ensures automated security against adversarial attacks that may interrupt the actual output of that AR space, which is mostly undiscovered. Predefined policy-based AR application security has limitations that may not be effective against entirely new and unpredictable attacks by malicious applications. For instance, an adversary may change the roadside traffic signs or block important information from the driver's view, leading the driver to make the wrong decision while driving. A self-learning technique for adaptive policies against malicious behavior is required to avoid such unexpected scenarios. AI-based reinforcement learning combined with fog computing plays a crucial role in developing defensive methods to filter out such potential attacks.

10.3.1 Categories of Attacks on Deep Learning Models

Proper investigation of attacks on deep learning-based models is required to deploy them in metaverse applications. Those attacks can be initially categorized into four types: model extraction attack, model inversion attack, poisoning attack, and adversarial attack. Research on those attacks for the security and privacy of the deep learning model has been increasing exponentially since 2014 [11].

10.3.1.1 Model Extraction Attack

This attack involves stealing the trained model without proper information regarding the algorithms and training data. The attackers duplicate the original model architecture by using APIs. The APIs throw input and associated query and receive output based on the query. Based on the output, attackers infer the model architecture and parameters and duplicate the model. The duplicated model is not only used to attack the privacy and confidentiality of the original model but also to conduct multiple attacks by developing a white box model [12].

10.3.1.2 Model Inversion Attack

Backpropagation is an important term in the domain of deep neural network training. After every iteration of the learning process, the parameters, including weights and biases, are updated based on the measurement of the cost function. An inverse information flow happens in that backpropagation. Attackers infer the training data during the backpropagation while training in model inversion attacks. This type of attack can be categorized into two sections: membership inference attack (MIA) and property inference attack (PIA). MIA is used to identify the presence of any specific record in the training data, whereas PIA is used to obtain any statistical properties in the data.

10.3.1.3 Poisoning Attack

A poisoning attack is generally the intervention of training data for deep neural network-based models. The goal is to degrade the model performance by polluting noise in the training data and control the model's behavior based on the inclusion of noise. This attack mainly controls the decision boundary essential to distinguish between target classes. The attackers may not only have access to full information of the data but also have the model architecture, parameters, hyperparameters, algorithms, and so on. Adversaries may temper the data by mislabeling them, which causes the wrong prediction for a given specific input.

10.3.1.4 Adversarial Attack

The main goal of this kind of attack is to figure out the weaknesses of the model by carefully observing the model's performance on the training set. For any misclassification sample, the attackers try to find out the reason for misclassification by the original model and exploit the input training data with the gained knowledge. This attack generally uses perturbation with the standard samples for training time and creates adversarial samples by observing the predictions from the trained model.

10.3.2 Adaptive Fog-based Output Security in AR/VR Space

Designing an intervention tool that deploys a reinforcement learning algorithm to detect the adversarial attacks or behavior and filters out those attacks before reaching the user's display in HMDs can secure the output of AR spaces. **Figure 10.2** depicts the pipeline of integrating reinforcement learning for the output security of AR systems.

- **Reinforcement Learning:** This method utilizes a trial-and-error learning mechanism, which mainly involves observing the surroundings and taking actions based on the collected rewards. In the security domain for the output of AR/VR applications, this method observes dynamic environments, learns the difference between malicious and normal behavior, and acts based on its learned algorithm. The advantages of the RL algorithm outperform the policy-based security system due to its ability to adjust rules based on its reward function. Compared to the supervised learning mechanism, the RL algorithm doesn't require labeled data to learn the behavior.
- **Fog Computing:** Utilizing an RL algorithm before visualizing any dynamic scene in AR applications requires a low-latency communication protocol framework. A fog node can be useful to provide relevant knowledge about the characteristics of the surroundings to the model to conduct the training procedure offline utilizing cloud servers. With the help of a local fog node, an AR device can take the information from a learned RL model and enhance the learning process by providing the appropriate feedback.
- **Policy optimization from learning algorithm:** In reinforcement learning, an agent learns to take optimum actions to evolve from one state to another in various dynamic environments based on the rewards received. The reward measures the agent's effectiveness in reducing

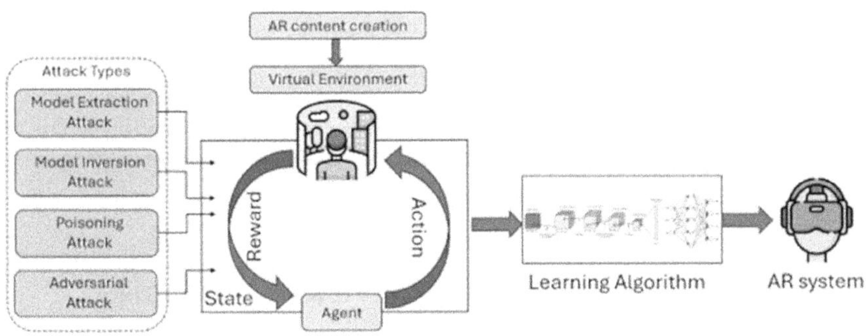

Figure 10.2 Pipeline of integrating the reinforcement learning for the output security of AR systems.

the distractions caused by AR content. Let us consider a function $f: C \rightarrow C'$, where the bijective function converts the original holograms to a policy-modified version. The reward at a time t would be,

$$R_t = \sum_{\substack{O' \in C' \\ R \in R}} \left(\pm . Dist_{XY}\left(R, O'\right) -^2 . Diff_Z\left(R, O'\right)\right) -^3 \sum_{\substack{O \in C, O' \in C' \\ f(O)=O'}} Dist\left(O, O'\right) \tag{10.1}$$

Here,

$$Dist_{XY}\left(R, O'\right) = \left|R_x - O'_x\right| + \left|R_y - O'_y\right| \tag{10.2}$$

$$Diff_Z\left(R, O'\right) = R_z - O'_z \tag{10.3}$$

$$Dist\left(O, O'\right) = \left|\right| Pos\left(O\right) - Pos\left(O'\right)\left|\right|_2 \tag{10.4}$$

Increment of distance $Dist_{XY}\left(R, O'\right)$ in XY-plane signifies that there is no obstruction from O' in the physical world, even after the adaptation of new policies. Penalizing the large z-plane ($Diff_Z\left(R, O'\right)$) is used to avoid the placement of O'infront of real-world objects. $Dist\left(O, O'\right)$ is the functionality of reward function that encourages the O' to be placed near the original holograms. Finally, α, β, and γ indicate the weights of these requirements.

According to the Markov Decision Process (MDP), the environment can be stated as (S, A, π, R), where $s \in S$ refers to the current state, $a \in \mathrm{A}$ represents the action taken by the agent, $\mathrm{R}\ (s, a) : SXA$ is the received reward, and the probability of transition from the old state s to new state s' is $\pi\left(s' \mid s, a\right) : SXAXS \rightarrow \left[0,1\right]$. The primary focus of the agent is to find an optimal policy π_θ based on the conducted actions and received rewards.

Here, $\theta^* = \arg\max\left(\theta\right) U\left(\theta\right) = \arg\max\left(\theta\right) \mathrm{E}\left[\sum_{t=1}^{H} R\left(S_t, a_t\right) \mid \pi_\theta\right]$ (10.5)

Here, $U\left(\theta\right)$ represents the gradient descent that is used to update the parameters by calculating loss, and H quantifies the future steps that may be required for the agents to get the optimum policy.

10.4 BIOMETRIC SECURITY IN AR/VR APPLICATIONS

User identification is required when using VR or AR applications such as gaming, rehabilitation and therapy, and e-commerce business. However, the traditional identification process mostly relies on choosing or entering a username and inputting unique PINs or passwords associated with that username through hand-held controllers or a virtual keyboard that provides users a certain level of discomfort of using such devices as well as raises the privacy and security concerns of the system. Thus, there is a clear need for ongoing research on the security in AR/VR applications.

10.4.1 Behavioral Biometrics

10.4.1.1 User Identification and Verification in VR

The traditional methods of user identification in VR (PINS, passwords) are slow in use and cause shoulder suffering up to 18% of authentication attempts. There is a clear need to conduct research on developing user-friendly authentication processes in VR applications that will reduce the disadvantages of using traditional authentication approaches. Measuring spatial relations between the controllers and HMD movement based on body motion analysis could be an option for the user identification process shown in [13]. Another approach could be the detection of the pattern of head movement by HMD [14]. 3D passwords based on the conjunction of gaze and head movements could also be employed as user authentication. Similarly, trucking hand motions while conducting a traditional identification approach could be used as a secondary form of biometrics.

10.4.1.2 Task-Based Biometrics

Task-driven biometrics is a subsection of behavioral biometrics. Collected data from the given manual task is used to measure unique characteristics for developing such an authentication process. To authenticate a driver's identification, some driving tasks can be employed [15] even if a single button press carries unique features that can be used for behavior characterization [16]. Eye tracking and gaze monitoring for specific tasks can also be used for such behavioral authentication processes. Moreover, the movement of balls, such as grabbing, and throwing from one position to another through hand-held controllers or HMD, can produce a higher degree of individual behavior that can be used for user authentication systems [17].

10.4.1.3 Illustrative Example

Liebers et al. [18] described both the authentication process and the task-driven biometrics in VR. The virtual scenarios have been developed for the Bowling and Archery tasks for user identification. In the bowling task, the features can be captured in three phases: 1. observing the bowl grabbing style and pressing motion on the grip button, 2. capturing swing motion as long as the ball is on the grip, and 3. observing hand movement after the ball is released. On the other hand, the goal of the archery task is to hit the bull's eye with the virtual arrow. Like the bowling task, the behavior characterization through the archery task can be conducted by maintaining several phases: (1) the process of attaching the arrow to the bow, (2) the style of grabbing strings along with the arrow, (3) the duration of pulling the string before releasing the arrow. and (4) the hand and neck movement until the arrow hits an object.

10.4.2 Human Visual System

Since VR is used in diverse domains such as gaming, healthcare, e-commerce, and the military, sensitive information can be retrieved by the head-mounted display (HMD). The premium version of applications requires purchasing most of the time, and credit cards or other forms of transaction information need to be entered by the hand-held remote control [19]. Besides, the training of the pilot for top secret aircraft is conducted with the VR simulator [20]. However, the VR system exposes the controller, and the attacker can intentionally observe the input type and infer sensitive information. The HMD, used in VR, covers the human eye. If the authentication process is developed based on the human visual system (HVS), the sensitive data can be secured from the risk of unauthorized access. Though eye gaze biometrics have been found in laptops and smartphone applications, the inclusion of this visual-based authentication process in VR is still challenging due to its high error rate and instability of accuracy [21–24]. Researchers are investigating beyond eyeball movement, diving into associated factors such as surrounding eye nerves, eyelid, and extraocular muscles for developing a visual-based user authentication process in VR as the anatomy of the HVS exhibits variation from person to person [25].

Physiological HVS features are collected by utilizing the EOG [26] signal acquisition system. Table 10.1. outlines the HVS features that need to be collected by electrodes in HMDs. Five electrodes are required to be placed close to the eye to capture the HVS activities. Among them, two electrodes are dedicated to capturing the HVS features from the outer edges of the eyes, two for the right eye, and the remaining one serves the purpose of capturing signals from the forehead as a reference. **Figure 10.3** depicts the appropriate position for electrode placement to capture the EOG signal.

10.4.3 Bio Sketch Security

Kinesiological movements in VR space can be utilized to identify users because people are kinesiologically unique. Due to the invention of modern VR devices, HMDs can track and record head movements by hardware sensors and monitor eye gaze movement with an eye tracker installed

Table 10.1 List of HVS features and associated electrodes

Index	*Features*	*Categories*	*Responsible Electrodes*
1	Rate of eyelid opening	Physiological	E1, E2
2	Rate of eyelid closing	Physiological	E1, E2
3	Eyelid stretch extend	Physiological	E1, E2
4	Initiation of fixation	Behavioral	E1, E2, E3, E4
5	Duration of fixation	Behavioral	E1, E2, E3, E4
6	Fixation focal point	Behavioral	E1, E2, E3, E4
7	Degree of upward rotation distance	Physiological	E1, E2
8	Degree of downward rotation distance	Physiological	E1, E2
9	Degree of rightward rotation distance	Physiological	E3, E4
10	Degree of leftward rotation distance	Physiological	E3, E4
11	Onset of saccade	Behavioral	E1, E2, E3, E4
12	Duration of saccade	Behavioral	E1, E2, E3, E4
13	Location of saccade	Behavioral	E1, E2, E3, E4

Note: E1, E2, E3, and E4 refer to electrode1, electrode 2, electrode 3, and electrode 4 respectively in Head-mounted display (HMD)

inside HMDs. Besides, the hand controller can be used to log users' hand movements. A VR space can be created, and a specified set of tasks can be conducted by the users. Based on the collected data of head, eye, and hand movements by the sensors in VR devices, a machine learning model can be used to classify user identification. Such a method mitigates the limitations of conventional user identification and authentication, such as PIN or password-based methods, by providing security against real potential observational attackers. **Figure 10.4** shows an illustration of the steps authentication process using the unique kinesiology features.

Creation of a VR environment: To conduct the task-driven kinesiology feature-based identification process, the creation of a virtual environment where the users can perform the defined tasks is required. Unity3D software can be used to do this task. Various types of environments and diverse tasks can be defined to observe the kinesiology movements. K. Lebeck, F. Roesner, and T. Kohno and I. Olade, C. Fleming, and H.-N. Liang [27, 28] created such an environment with a wooden stand and three horizontal poles attached to it, where each horizontal pole carries one cubic container and one cylindrical container. Some cubes and balls can be placed beside the stand. The

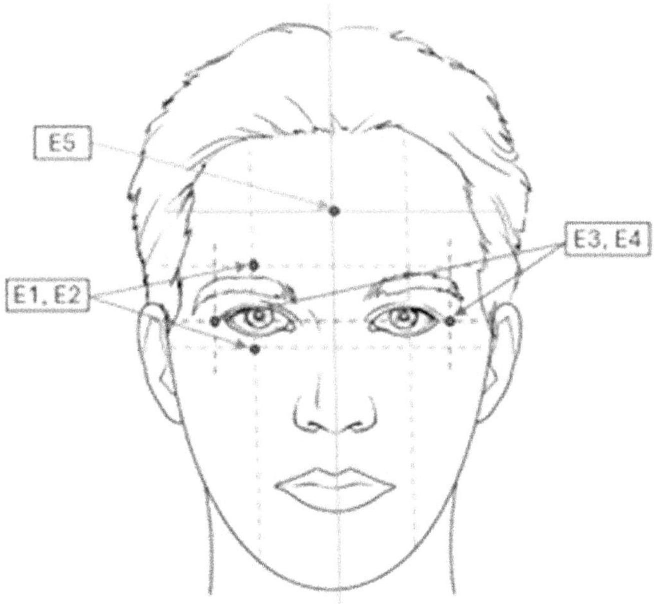

Figure 10.3 Electrode placement for EOG signal acquisition.

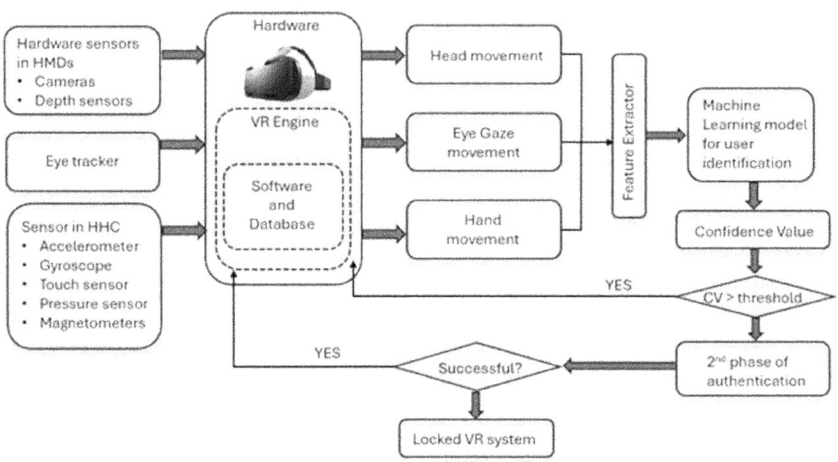

Figure 10.4 Two-step authentication process using the unique kinesiology features.

task for the participants should grab those cubes and balls, transport them through the path, and drop them according to the container shapes. While doing these simple sets of tasks, the head movement with a possible 6 degrees of freedom can be recorded by HMDs, eye gaze movement can be recorded by the eye tracker in HMDs, and hand movement with 6 possible degrees of freedom can be recorded by HHC (Hand-held controller) to identify the uniqueness of the kinesiology.

The task-performing speed can vary from person to person. However, the system captures the movement vectors with a fixed frequency (For instance, 25 movement vectors per second). Thus, the number of captured movement vectors changes with the speed of performing tasks for the same identical task. So, the data must be resampled to maintain the consistency of movement vectors that will avoid the risk of degradation of model performance. The process of resampling the data has been given below.

A session S can be defined as a set of tasks T, as described in Equation (10.6).

$$S = \{t_1, t_2, \ldots, t_6\}, where\, the\, number\, of\, task\, is\, 6 \tag{10.6}$$

Task T consists of a set of movement vectors m, as shown in Equation (10.7)

$$T = \{m_1, m_2, \ldots, m_n\}, where\ the\ number\ of\ movement\ vectors\ is\ n \tag{10.7}$$

The median value D, representing the size of movement vectors |T| for each task category $\{t_1, t_2, \ldots, t_6\}$ throughout all sessions S is calculated in the following manner: For every Task category Tx (where $1 \leq x \leq 6$),

$$D_x = Median\left(|T_x|_1, \ldots, |T_x|_n\right) \tag{10.8}$$

The median or sample size D_x was calculated for each Task type T_x using Equation (10.8). The movement count |T| for each Task was then adjusted through resampling to match the respective D_x count. This ensured that by the end of the resampling process, all task groups had movement vectors of equal cardinality.

10.5 BLOCKCHAIN IN AR/VR SPACE

Privacy and security concerns are the most crucial part of building the metaverse ecosystem, and blockchain is one of the core infrastructures in that ecosystem that provides security to other technologies/applications in

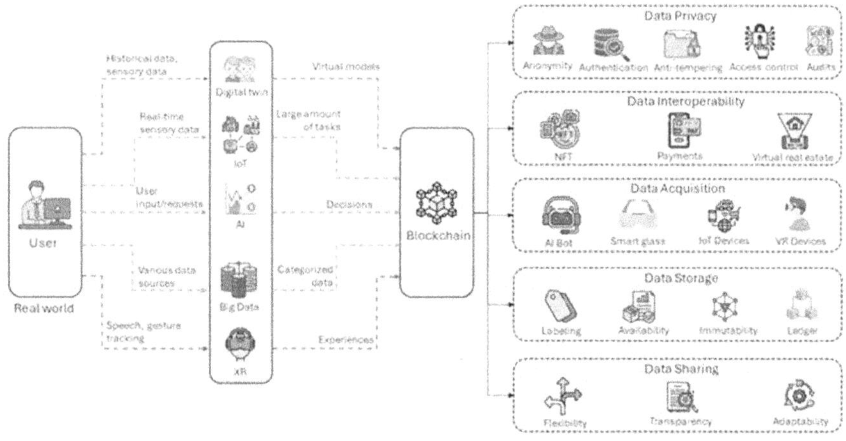

Figure 10.5 Role of blockchain in metaverse ecosystem.

the metaverse. For instance, blockchain data block architecture and hash function techniques offer high-speed data transmission security. Blockchain provides a decentralized architecture that reduces the demerits of the centralization structure, such as data confidentiality risks. As shown in **Figure 10.6**, blockchain provides a decentralized architecture that provides data privacy by utilizing anonymity, access control mechanisms, data interoperability using NFT, secured credit payments, cloud-based fast and reliable data acquisition systems from devices like smart glasses, and VR HMDs, highly protected data storage capability, and data sharing with transparency, adaptability, and flexibility. **Figure 10.7** depicts the layered architecture of blockchain. Like other elements of the metaverse ecosystem, AR/VR also demands highly secured space, and the inclusion of blockchain can meet this requirement.

Data security in multi-user space in AR/VR: Multiple users can communicate with each other in virtual space leveraging the AR/VR technology. The communication requires not only high-speed bandwidth but also secure storage for a large amount of generated data. The centralized architecture may be overloaded due to this multiple asynchronous communication. As shown in **Figure 10.7**, blockchain has a transmission and data layer that enables users to have a decentralized P2P transmission and verification mechanism. It also provides data security by utilizing a hash algorithm and chain structure.

Transmission security in AR/VR-based applications: Law enforcement authorities and militaries emphasize cybersecurity in highly confidential applications. In the metaverse, the usages of AR/VR in those applications will be increased. Blockchain can enhance cybersecurity

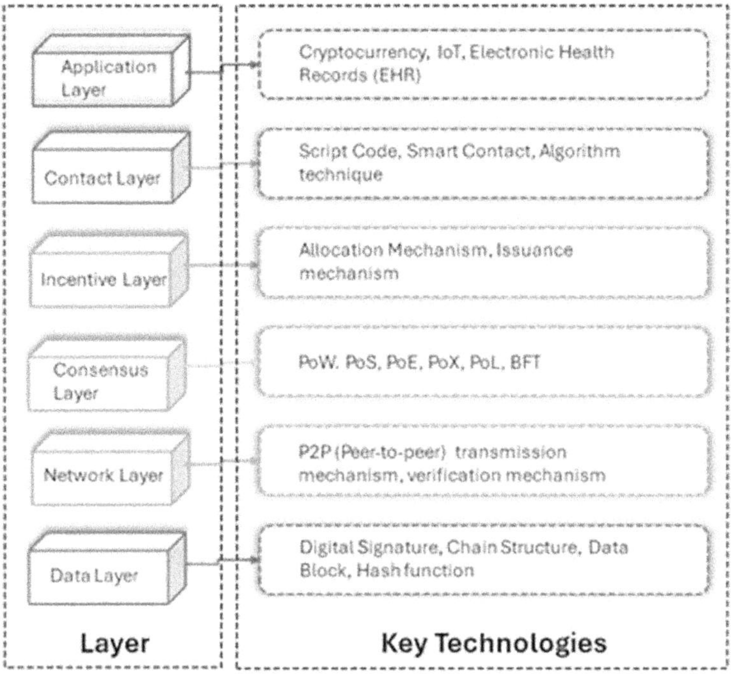

Figure 10.6 Layered architecture of blockchain.

in AR/VR applications by providing low latency and encrypted block-based transmission protocol.

Security in digital assets: The metaverse introduces the term "digital asset," which mainly refers to the virtual hidden property meant to be traded. Trading won't be conducted via cash or credit card-based methods; rather, virtual currencies will be introduced in the metaverse. Cryptocurrency is the most popular example of a digital asset empowered with blockchain. The price of the major cryptocurrencies is rising daily due to the trust of a wide range of users in them.

To solve the credit transaction problem caused by frauds or hackers, blockchain offers a consensus layer that includes various consensus like PoW, PoS, and PoE. One of the most popular cryptocurrencies, "Ethereum," uses the Proof of Work (PoW) consensus algorithm. In PoW, the network nodes engage in solving robust puzzles. The node can only become the block prosper which can successfully solve the puzzle and receive the rewards. This process involves exhaustively searching for a specific nonce value that, when combined with other data, results in a cryptographic hash meeting predefined condition.

$$Hash(x, nonce) \leq target \quad .$$

Here, "x" represents the header of the previous block, while "target" denotes the difficulty level of the current block and miners search for a solution, referred to as the "nonce." Currently, the exchanges of most cryptocurrencies happen within a centralized architecture. Though there are some benefits of centralized transactions, such as low latency, security issues due to price manipulation will always be a concern. Since the blockchain provides the decentralized architecture of transactions, the cryptocurrency transaction in the metaverse applications such as AR/VR spaces will be empowered by blockchain. That will help to build a reliable ecosystem in the metaverse.

10.5 FUTURE OPPORTUNITIES

Modern technologies have contributed significantly to solving metaverse applications' security and privacy issues. However, many areas of improvement remain that need to be considered before implementing a reliable metaverse ecosystem.

Though the architecture of a metaverse ecosystem is decentralized in nature, the stockholder of any metaverse ecosystem will have the logs of the user data, and that may lead to threats to the confidentiality and integrity of user privacy. Adequate research is needed to be undertaken to solve this issue.

Psychological manipulation can occur through social engineering, which may cause breaches in information flow. Due to this, law enforcement authorities may face difficulties identifying criminals in the metaverse space. Comprehensive studies on detecting the attempt of psychological manipulation in the metaverse should be conducted.

Miners solve the hash puzzles to mine the blocks that are used for cryptocurrency transactions, which is a fundamental part of trading in the metaverse securely. However, this process requires a lot of computing power and electricity. Though quantum computing is introduced to mitigate the issue of huge computing power, there is a lot of scope for research to decrease dependency on high computing power and electricity for using blockchain architecture.

Records of Images of human faces, body movements, and speech may pose a great security threat in the metaverse ecosystem and compromise biometric security in AR/VR space. Storing the logs of those records with highly secured architecture is another important field of research.

10.6 CONCLUSION

Like the IoT is the blessing of today's invention, the metaverse ecosystem will be the groundbreaking invention of next-generation technology. Though the metaverse will come with an enormous number of blessings from modern technologies, some serious security and privacy threats need to be considered. In this chapter, we comprehensively discussed the architecture of the metaverse focused on AR/VR space, possible adversarial attacks and threats, and the attack prevention mechanism based on AI, behavioral security, and Blockchain. Additionally, some essential open challenges and issues have been addressed to employ the metaverse ecosystem and the future research direction related to those constraints. This chapter lays a fundamental foundation that helps to understand the components of the metaverse focused on AR/VR space with associated privacy and security concerns.

REFERENCES

1. D. Grider and M. Maximo, "The metaverse: Web3.0 virtual cloud economies," accessed Nov. 1, 2021. [Online]. Available: https://grayscale.com/wp-content/uploads/2021/11/Grayscale_Metaverse_Report_Nov2021.pdf.
2. C. Jaynes, W. B. Seales, K. Calvert, Z. Fei, and J. Griffioen. 2003. "The Metaverse: A networked collection of inexpensive, self-configuring, immersive environments," in *Proceedings of the workshop on Virtual environments 2003 (EGVE '03)*. Association for Computing Machinery, New York, NY, USA, 115–124. https://doi.org/10.1145/769953.769967
3. Y. Wu, K. Zhang, and Y. Zhang, "Digital twin networks: A survey," *IEEE Internet of Things Journal*, vol. 8, no. 18, pp. 13789–13804, Sep. 2021.
4. H. Du, D. Niyato, J. Kang, D. I. Kim, and C. Miao, "Optimal targeted advertising strategy for secure wireless edge metaverse," *arXiv:2111.00511*, 2021.
5. S. Vural, D. Wei, and K. Moessner, "Survey of experimental evaluation studies for wireless mesh network deployments in urban areas towards ubiquitous Internet," *IEEE Communications Surveys & Tutorials*, vol. 15, no. 1, pp. 223–239, 1st Quart., 2013.
6. T. Huynh-The, Q.-V. Pham, X.-Q. Pham, T. T. Nguyen, Z. Han, and D.-S. Kim, "Artificial intelligence for the metaverse: A survey," *arXiv:2202.10336*, 2022.
7. C. T. Nguyen, D. T. Hoang, D. N. Nguyen, and E. Dutkiewicz, "Metachain: A novel blockchain-based framework for metaverse applications," in *Proceedings of the 2022 IEEE 95th Vehicular Technology Conference (VTC2022-Spring)*, Helsinki, Finland, Jun. 19–22, 2022, pp. 1–5. doi: 10.1109/VTC2022-Spring54318.2022.9860983
8. R. Menzies et al., "An objective measure for the visual fidelity of virtual reality and the risks of falls in a virtual environment," *Virtual Reality*, vol. 20, no. 3, pp. 173–181, 2016.

9. C. Kai, H. Zhou, Y. Yi, and W. Huang, "Collaborative cloud-edge task offloading in mobile-edge computing networks with limited communication capability," *IEEE Transactions on Cognitive Communications and Networks*, vol. 7, no. 2, pp. 624–634, Jun. 2021.
10. Goldman Sachs, "Virtual & augmented reality," Jan. 2016. [Online]. Available: http://www.goldmansachs.com/our-thinking/pages/technology-driving-innovation-folder/virtual-and-augmented-reality/report.pdf.
11. Y. He et al., "Towards privacy and security of deep learning systems: A survey," *arXiv preprint arXiv:1911.12562*, 2019.
12. N. Papernot, P. D. McDaniel, I. J. Goodfellow, S. Jha, Z. B. Celik, and A. Swami, "Practical black-box attacks against machine learning," in *Proceedings of the 2017 ACM on Asia Conference on Computer and Communications Security (AsiaCCS)*, United Arab Emirates, Abu Dhabi, Apr. 2–6, 2017, pp. 506–519.
13. K. Pfeufer, M. J. Geiger, S. Prange, L. Mecke, D. Buschek, and F. Alt, "Behavioural biometrics in VR: Identifying people from body motion and relations in virtual reality," in *Proceedings of the 2019 CHI Conference on Human Factors in Computing Systems* (Glasgow, Scotland, UK) (CHI '19), Association for Computing Machinery, New York, NY, pp. 1–12, 2019. [Online]. Available: https://doi.org/10.1145/3290605.3300340.
14. T. Mustafa, R. Matovu, A. Serwadda, and N. Muirhead, "Unsure how to authenticate on your VR headset? Come on, use your head!," in *Proceedings of the Fourth ACM International Workshop on Security and Privacy Analytics* (Tempe, AZ, USA) (IWSPA '18), Association for Computing Machinery, New York, NY, pp. 23–30, 2018. [Online]. Available: https://doi.org/10.1145/3180445.3180450.
15. K. Igarashi, C. Miyajima, K. Itou, K. Takeda, F. Itakura, and H. Abut, "Biometric identification using driving behavioral signals," in *Proceedings of the 2005 IEEE Intelligent Transportation Systems Conference*, Vienna, Austria, Sep. 16, 2005. doi: 10.1109/ITSC.2005.1520171
16. H. Pohl, M. Krause, and M. Rohs, "One-button recognizer: Exploiting button pressing behavior for user differentiation," in *Proceedings of the 2015 ACM International Joint Conference on Pervasive and Ubiquitous Computing* (Osaka, Japan) (UbiComp '15), Association for Computing Machinery, New York, NY, pp. 403–407, 2015. [Online]. Available: https://doi.org/10.1145/2750858.2804270.
17. I. Olade, C. Fleming, and H.-N. Liang, "BioMove: Biometric user identification from human kinesiological movements for virtual reality systems," *Sensors*, vol. 20, no. 10, p. 2944, 2020.
18. J. Liebers, M. Abdelaziz, L. Mecke, A. Saad, J. Auda, U. Gruenefeld, F. Alt, and S. Schneegass, "Understanding user identification in virtual reality through behavioral biometrics and the effect of body normalization," in *Proc. 2021 CHI Conf. on Human Factors in Computing Systems (CHI '21)*, New York, NY, USA: Association for Computing Machinery, May 2021, Art. no. 517, pp. 1–11. doi: 10.1145/3411764.3445528
19. J. Liebers et al., "Rift store: VR games, apps & more," [Online]. Available: https://www.oculus.com/experiences/rift/.
20. V. R. Society, "Virtual reality air force training," [Online]. Available: https://www.vrs.org.uk/virtual-reality-military/air-force-training.html.

21. M. Porta, S. Ricotti, and C. J. Perez, "Emotional e-learning through eye tracking," in *Proc. 2012 IEEE Global Engineering Education Conference (EDUCON)*, Marrakech, Morocco, Apr. 17–20, 2012, pp. 1–6. doi: 10.1109/EDUCON.2012.6201145
22. I. Sluganovic, M. Roeschlin, K. B. Rasmussen, and I. Martinovic, "Using reflexive eye movements for fast challenge-response authentication," in *Proc. 2016 ACM SIGSAC Conf. on Computer and Communications Security (CCS '16)*, New York, NY, USA, 2016, pp. 1056–1067. doi: 10.1145/2976749.2978311.
23. C. Song, A. Wang, K. Ren, and W. Xu, "Eyeveri: A secure and usable approach for smartphone user authentication," in *Proc. IEEE INFOCOM 2016 – The 35th Annual IEEE International Conference on Computer Communications*, San Francisco, CA, USA, Apr. 10–14, 2016, pp. 1–9. doi: 10.1109/INFOCOM.2016.7524367.
24. Y. Zhang, W. Hu, W. Xu, C. T. Chou, and J. Hu, "Continuous authentication using eye movement response of implicit visual stimuli," *Proceedings of the ACM on Interactive, Mobile, Wearable and Ubiquitous Technologies*, vol. 1, no. 4, p. 177, 2018.
25. M. Joukal, "Anatomy of the human visual pathway," in *Homonymous Visual Field Defects*, K. Skorkovská, Ed., Cham, Switzerland: Springer, Apr. 2017, pp. 1–16. doi: 10.1007/978-3-319-52284-5_1.
26. A. Bulling, J. A. Ward, H. Gellersen, and G. Troster, "Eye movement analysis for activity recognition using electrooculography," *IEEE Transactions on Pattern Analysis and Machine Intelligence*, vol. 33, no. 4, pp. 741–753, 2010.
27. K. Lebeck, F. Roesner, and T. Kohno, "Securing augmented reality output," in *Proc. 2018 IEEE Symposium on Security and Privacy (SP)*, San Francisco, CA, USA, May 20–24, 2018. doi: 10.1109/SP.2018.00051.
28. I. Olade, C. Fleming, and H.-N. Liang, "BioMove: Biometric user identification from human kinesiological movements for virtual reality systems," *Sensors*, vol. 20, no. 10, p. 2944, 2020.

Part III

Metaverse Applications

Chapter 11

3D Holograms for Stroke Rehabilitation

Trent Boughner, Nashmi Alheraijy, Qais Altannak, Patrick Brooks, Luke Burbee, Zane Pautzke, and Fei Hu

11.1 INTRODUCTION

Mixed reality technology increasingly embeds itself into modern society. While virtual reality offers escapism and a sense of immersion into an entirely digital world, augmented reality meshes holographic content into the physical world. Both ends of the mixed reality spectrum enable developers to create applications for diverse use cases in industry. Companies have introduced a catalogue of devices to capitalize on this new frontier with the intention of permanently integrating mixed reality into every aspect of the lives of consumers. Microsoft introduced the HoloLens [10] in 2015 and followed with a sequel, Valve developed virtual reality headsets with augmented reality cameras, and Apple released the Apple Vision Pro in early 2024. With its purchase of the Oculus brand from Google, Meta entered the mixed reality market and invested billions of dollars into developing a shared mixed reality ecosystem titled the "Metaverse" that intends to permanently integrate digital overlays into everyday life. Despite the efforts of industry giants, the emerging technology remains relatively new and must be significantly enhanced to become mainstream, fulfill the use cases of various consumer demographics, and match the quality of the technology exhibited in science fiction media.

The impact of strokes on the brain and motor skills of the body cannot be understated. When a patient has a stroke, time is of the essence, as the impact that a stroke has on the brain worsens as the amount of time that passes before the patient receives treatment increases. Stroke victims must therefore travel to the nearest stroke-ready medical center and connect with a medical professional to receive immediate treatment. Utilizing various modes of transportation, especially planes and automotives, to move patients between medical centers often requires large sums of money and increases the carbon emissions released into the environment.

Introducing mixed reality into the medical field will revolutionize the process of stroke rehabilitation. When stroke patients do not have in-person access to a medical professional, communicating with a doctor via mixed reality can increase the speed at which they receive treatment by drastically

DOI: 10.1201/9781032663609-14

reducing the required amount of travel. This will decrease the expenses that the patient incurs and emissions that vehicles will release into the atmosphere. Since mixed reality simulates in-person interaction, the doctor will harness the ability to perform physical and cognitive evaluations on patients undergoing stroke rehabilitation despite not being in the same physical location.

To fully implement a mixed reality system that solves these issues, medical firms need to invest in a new infrastructure of therapeutic centers with a system of mixed reality devices and stroke rehabilitation equipment. Integrating such a system into the medical field will enable doctors to save and transform lives more efficiently.

11.2 DELIVERABLES

The chapter aims to create an integrated system of hardware and software applications that simulate in-person doctor's appointments within augmented reality. To complete the objective, the system must complete the following deliverables.

1) The system must complete the process of "Holoportation," which involves creating a three-dimensional holographic avatar of the patient within the doctor's perspective that visualizes the patient's movement in real time.
2) The system must support voice communication over a network between the doctor and the patient.
3) The system must enable medical professionals to perform physical and cognitive evaluations of patients undergoing stroke rehabilitation over a network by utilizing mixed reality and stroke rehabilitation equipment.

11.3 DESIGN PRINCIPLES

11.3.1 System Overview

The chapter employs two subsystems: 3D Capture and Communication and XR Game Development (as shown in Figure 11.1). The 3D capture portion implements the algorithm that scans the patient while he or she operates the treadmill and HoloLens and then creates a holographic 3D point cloud within the mixed reality space that updates in accordance with the patient's real-time movements. The communication portion of the subsystem integrates voice communication over a network between the doctor and the patient. The system transfers the holographic avatar into the HoloLens perspective of the doctor, who can then alter the size, position,

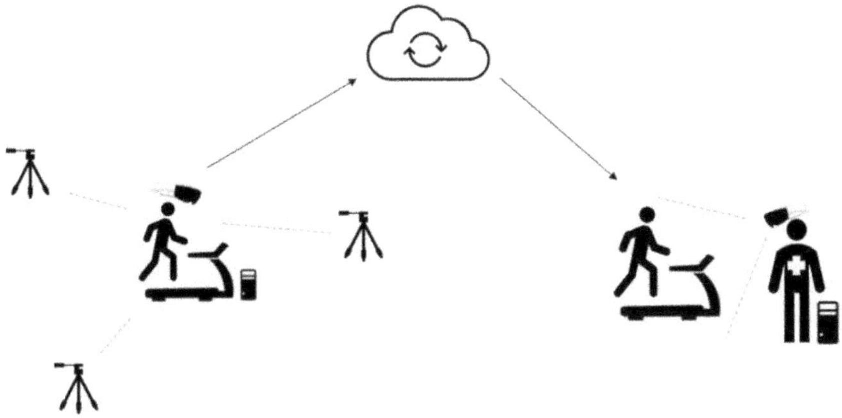

Figure 11.1 Outline of subsystems.

and orientation of the hologram as it updates. The XR game development subsystem includes mixed reality minigames that test the patient's coordination, balance, and cognition.

11.3.2 Key Design Decisions and Challenges

The chapter presented a unique and challenging opportunity for the team to gain experience in the world of mixed reality. The chapter expertly blended technical expertise with an artistic medium while presenting an innovative solution to stroke rehabilitation. The chapter granted the team a great deal of creative freedom by enabling them to design mixed reality video games that fulfill the business and medical use cases of industry professionals.

The senior design team consisted of three computer engineering majors and three electrical engineering majors from the University of Alabama's Electrical and Computer Engineering department. Despite the team's experience with hardware or its integration with software, the chapter primarily concerned the discipline of computer science and utilized programming languages such as C# rather than C or hardware description languages. Several of the electrical engineering majors had limited programming experience and required a considerable amount of ramp-up to develop scripts in C#. One of the electrical engineering majors joined the chapter at the beginning of the spring semester. While the team successfully gained development experience and completed the deliverables, the situation emphasized the importance of conducting proper training and developing detailed knowledge management systems in an organization. Figure 11.2 shows the outline of the subsystem and the team formation.

The experience granted each team member exposure to working on technical chapters with members of differing demographics, cultural

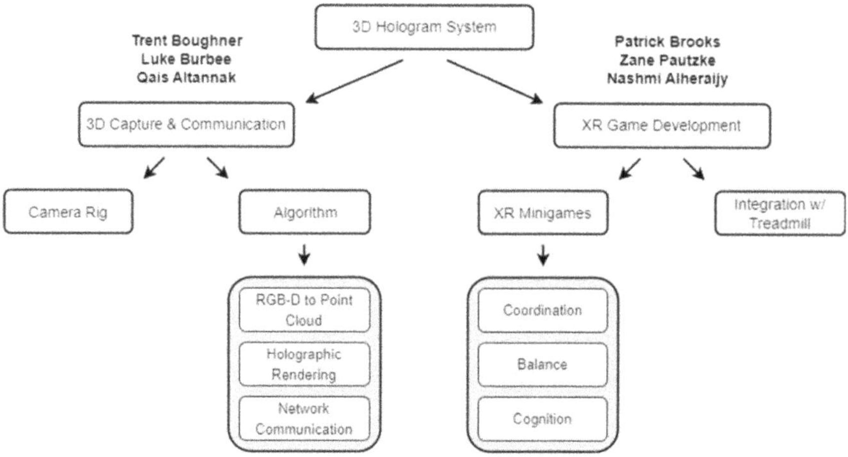

Figure 11.2 Integration of subsystems.

backgrounds, and subsequent communication styles. Dr. Hu, who has an electrical engineering background, and his Ph.D. students, who have computer science backgrounds, originated from China, while the senior design team members originated from the United States and Kuwait. Communication issues occasionally arose when the Ph.D. students verbally provided instructions for the technical deliverables of the chapter in a low-context manner to the American students, who traditionally operate with more high-context communication from superiors in the classroom and workplace. The low-context and more open-ended instructions, while encouraging the team to think critically, occasionally lacked clear direction and created some confusion regarding deliverables. Creating team meetings where each party elaborated extensively about design requirements and expectations resolved these issues. The group also created concrete, biweekly deadlines for deliverables in accordance with the chapter's schedule, which increased the team's efficiency, motivation, and morale.

The team chose to develop the software for the chapter within the Unity game engine [1–3]. Compared to competing engines such as Unreal Engine, which prioritizes the use cases of large AAA game development teams, Unity offers an environment that caters to smaller teams of independent developers. Microsoft's creation of the Mixed Reality Toolkit, a plugin for Unity that enables developers to develop and deploy Unity chapters to Microsoft HoloLens 2, increased support for the compatibility of devices such as Azure Kinect DK [1] with the Unity game engine. After extensive research, the team selected Unity for the development of the system.

For the chapter to run successfully, specific versions of each software component had to be utilized to ensure compatibility. When one developer updates a software component independently from the others, the developers of the other components often do not update their own products to support that newest version. The team therefore invested a considerable amount of time and effort in reading documentation and analyzing version histories to ensure the compatibility of all components within the system. For example, MRTK3 did not share compatibility with versions of Unity past 2021; the team therefore developed both subsystems with a version of Unity from 2021 so that the Unity chapters containing the avatar and minigames could successfully deploy to the HoloLens devices. The XR game development system also spent a considerable amount of time troubleshooting the process of importing and displaying treadmill data, including the patient's speed in meters per second, in the Unity environment.

The team faced several challenges concerning the accessibility of hardware components. Due to shortages of electronic components caused by the COVID-19 pandemic, the team did not receive the Microsoft HoloLens 2 devices until November 2022. While awaiting their arrival, the team gained experience developing basic chapters in mixed reality for the Valve HTC Vive Pro 2. Despite being a virtual reality headset, the device includes an augmented reality camera that translates the view of the physical world directly in front of the user into the virtual reality screen inside the headset. However, the most recent version of Valve's AR game development kit for the device contained significant bugs that prevented the SDK from working properly. Despite the situation, the team gained valuable experience developing scripts in C# and working with Unity engine and numerous third-party Unity plugins.

The team faced another challenge related to documentation. Due to Microsoft's strict vetting over devices such as the Microsoft HoloLens and Azure Kinect DK [16], very few individuals have the privilege of utilizing them along with many of the software components that the chapter utilizes. Due to the relatively small userbase for the devices and software components such as MRTK3, very little relevant documentation exists outside of what Microsoft created for its own devices. The scarce amount of documentation increased the difficulty of integrating components, debugging scripts, etc. Despite the lack of textual documentation, numerous independent developers have published open-source content on platforms such as YouTube. The team referenced these resources to troubleshoot bits and pieces of the system and gain inspiration for new and improved methods of completing the system's deliverables.

Another major challenge involved combining the work of both subsystems into one Unity chapter. The team ultimately could not create one singular chapter with the functionality of both subsystems due to issues integrating certain software components. However, the team discovered that completing the deliverables did not require such integration and split the system into two communication channels. Each subsystem created one of the two. The 3D capture subsystem's channel, which controlled the doctor's perspective, transferred the point cloud data from the Kinects into Unity via a wired connection to an engineering laptop, which deployed the point cloud into the doctor's HoloLens. The XR game development subsystem's channel transferred data from the treadmill to the desktop PC via Ethernet while the patient operated the treadmill and then deployed the data into the patient's HoloLens. Having each subsystem work on a different computer streamlined the development process by allowed both subsystems to work in the laboratory simultaneously.

Once the team obtained every device required to build the full system, the most difficult challenge of the chapter involved the 3D capture and communication subsystem's creation of the 3D avatar. The team essentially had two options: create or find open-source software that uses machine learning, a single Kinect, and the Kinect's SDK to generate a full 3D point cloud of the patient, or position three Kinects around the patient in a 360-degree manner and mesh the three corresponding point clouds into one object within the Unity scene. Due to the chapter's nine-month time constraint coupled with the extensive time commitment required to learn and develop artificial intelligence, the team opted for the former option. The team spent several months incorporating numerous open-source algorithms into Unity chapters and modifying them to complete the deliverables; many of the algorithms either could not be fully integrated with every device in the system or complete the deliverables without extensive additional development. The team finally found "Azure Kinect Examples for Unity," a Unity package with a body-tracking algorithm that utilized the Azure Kinect DK [8] cameras to create point clouds of the patient and deploy them to the Microsoft HoloLens 2 via MRTK3. Each of the three cameras created a different point cloud, and the team manually meshed the three point clouds into one to create the best quality model possible and complete the final deliverable.

11.3.3 Software Requirements

Key software components and the corresponding versions include Azure Kinect SDK 1.4.1, Azure Kinect Body Tracking SDK 1.1.2, Azure Kinect Examples for Unity, Unity Hub 3.4.1, Unity 2021.2.7f1, High-Definition Render Pipeline (HDRP), Mixed Reality Feature Tool (MRTK 3) [5], Mixed Reality OpenXR [6] Plugin, and Photon Voice 2. The chapter utilizes version 2021.2.7f1 of the Unity editor to create chapters that share compatibility with Mixed Reality Toolkit 3.

The team purchased the Azure Kinect Examples for Unity package from the Unity asset store. The package supports Azure Kinect DK, Kinect-v2 sensor, and Phone-Pro LiDAR sensors and includes a body-tracking algorithm that creates point clouds of humans and updates them in real time based on their movement. It contains example Unity scenes and C# scripts necessary to transfer the Azure Kinect point cloud data into Unity and render it as a point cloud. It also contains network transmission functionality. The team modified the source code and Unity chapter to integrate multiple cameras into the system. The HDRP Unity package increases the graphical quality of holograms within the Unity scene. The package is necessary to use and configure Azure Kinect Examples for Unity within a chapter.

Microsoft developed the MRTK3 Unity plugin to enable mixed reality Unity chapters to be developed and deployed to the Microsoft HoloLens 2. Unity contains a wizard that must be used to add MRTK 3 to Unity Chapter. To use the tool, run ("MixedRealityFeatureTool.exe"), specify the root of the Unity chapter, and select desired packages. The team installed every feature using MRTK3.

The Mixed Reality OpenXR [11] plugin is available for installation from the Mixed Reality feature tool [12] and must be configured into the chapter to target the Microsoft HoloLens 2 upon building the chapter.

The Photon Voice 2 Unity package implements voice communication with real-time audio. Industry developers often use it to create multiplayer games.

11.3.4 3D Capture and Communication – Point Cloud Creation

These components render RGB-D data from the Azure Kinects as a 3D point cloud in a Unity scene. The Kinect4Azure Unity package increases the graphical quality of holograms within the Unity scene. The package is necessary to use and configure Azure Kinect Examples for Unity within a chapter. The Kinect4AzureInterface script allows the system to retrieve data from a Kinect. A PointCloudColorMap and PointCloudVertexMap texture must be specified to render a point cloud. Each Kinect requires its own Kinect4Azure instance, which must be placed as a child to the corresponding instance of KinectController. The KinectController object works within the Kinect Manager script to pull necessary data from the Kinects. Possible options for data selection include depth, color, infrared, pose, and body frames. Each Kinect requires a PointCloud object within the ColorPointCloud asset template. It also includes a Box Collider and Object Manipulator to make the point cloud interactable. The ColorPointCloud asset template consists of the PointCloudColorMap and PointCloudVertexMap textures and needs to be attached to the PointCloud object. Each map must correspond to the correct desired Kinect.

11.3.5 3D Capture and Communication – Network Transmission

These components send and receive Kinect data across a network. The KinectNetServer object works with the Kinect Net Server script. It must be given a sensor index and base port. The object will set up a range of ports listening for requests of Kinect data. The system utilizes one port for each type of data; specifying sensor index 0 and base port of 11000 creates a DepthServer on 127:0:0:1:11108 that only listens to requests for and responds with depth data. Each Kinect requires a server. The NetClient object, comparable to Kinect4Azure, acts as the interface between Unity and a remote Kinect. The IP address and base port of the server must be specified. To turn this data into a point cloud, PointCloudColorMap and PointCloudVertexMap, textures must be provided to write to.

11.3.6 3D Capture and Communication – Audio

Photon Voice 2 adds high-quality low-latency voice chat to Unity chapters. The Voice Connection and Recorder object works with numerous Unity Voice Client, Recorder, Connect and Join, and Mic Amplifier scripts. Together, the objects are responsible for interacting with the PC microphone, amplifying the audio, and connecting to a Photon Voice Server. The VoiceLogger object includes the Voice Logger script to provide debug information.

11.3.7 KineAssist – Mobility eXtreme (MX)

The KineAssist – MX [7,9] treadmill enables users to perform rehabilitation exercises related to walking and balance. The interface of such treadmill is shown in Figure 11.3. Exercises act as a form of treatment for various medical impairments concerning the neurological, cardiovascular, respiratory, muscular, or skeletal systems.

The treadmill directly connects to the host Alienware PC via wired Ethernet and transfers data via the Telnet communication protocol. As the patient operates the treadmill, the treadmill's software transmits data to the host PC, enabling Unity to interact with the data. The patient's lower body utilizes the treadmill as a controller for the mixed reality games, enabling the doctor to track metrics associated with balance, walking speed, etc. The treadmill includes a harness with three safety straps. The patient can adjust the height of the safety harness by placing his or her fingers within the red rings at the ends of the harness. If the patient loses balance, the treads will lock up and the harness will catch the patient. The green and red buttons on the computer terminal serve to start and stop the treadmill, respectively. The user can toggle a body weight supported (BWS) catch system on and

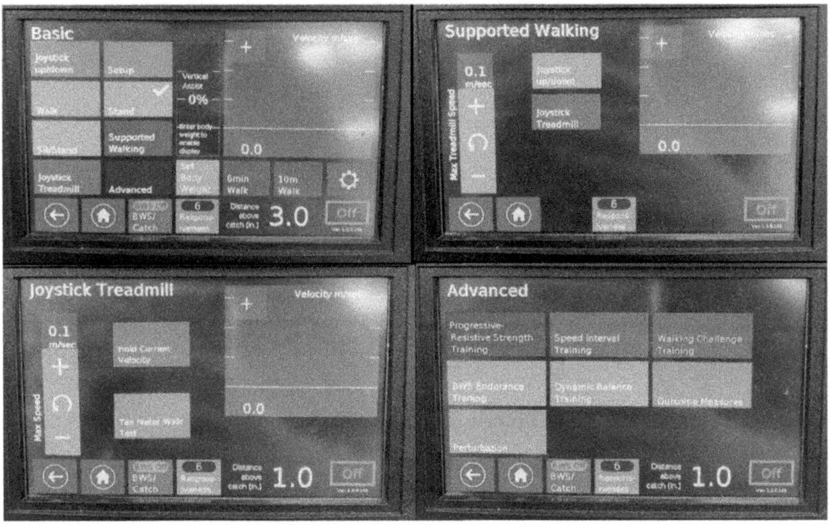

Figure 11.3 KineAssist Mobility eXtreme (MX) Computer Terminal User Interface.

off. The user can also adjust a vertical assist setting, responsiveness setting, and the distance above the catch in inches.

The computer terminal connected to the treadmill contains a user interface that displays several adjustable settings for alternate operating modes. The joystick up/down mode locks the treads and configures the joystick so that the user can vertically adjust the harness. The setup mode locks both the harness and the treads so that the user can more easily alter settings. The walk mode locks the harness in place and unlocks the treads so that the user can walk or run freely. The stand mode locks the treads and unlocks the harness to freely move vertically horizontally perpendicular to the direction of the treads. The mode tests the user's balance. The sit/stand mode unlocks both the treads and the harness but limits the total distance that patient can travel forward and backward on the treadmill. The supported walking mode enables the user to operate the treadmill with a manually set maximum speed. This mode can be combined with the joystick up/down to determine the method with which the user activates the treadmill at a constant speed. The joystick treadmill setting enables the user to set a maximum speed, hold current velocity, and perform a 10-meter walk test. The advanced modes include progressive-resistive strength training, speed interval training, walking challenge training, body weight supported (BWS) endurance training, dynamic balance training, outcome measures, and perturbation.

Figure 11.4 Azure Kinect DK.

11.3.8 Microsoft Azure Kinect DK (x3)

The Microsoft Azure Kinect DK [4] cameras contain software development kits that enable developers to create advanced computer vision and speech models using light detection and ranging (LiDAR) technology. The hardware of Azure Kinect is shown in Figure 11.4 and the user interface of the Azure Kinect DK is shown in Figure 11.5. The cameras utilize sensors embedded with artificial intelligence to produce RGBD images, which incorporate color and depth. The device includes a 7 microphone array, infrared emitters, RGB camera, depth camera, gyroscope, and accelerometer. Each camera has USB-C to USB-A data cable and a USB power supply. Connecting multiple cameras into a daisy-chain configuration requires a compatible 3.5 mm audio sync cable. The body tracking algorithm included within the Azure Kinect Examples for Unity open-source algorithm interfaces between the device's SDK and Unity. Additional specifications can be found at https://learn.microsoft.com/en-us/azure/kinect-dk/hardware-specification.

11.3.9 Microsoft HoloLens 2 (x2)

The Microsoft HoloLens 2 is an augmented reality headset that renders interactable 3D holograms in the real world. The system supports hand-tracking and eye-tracking, enabling the user to create and interact with holographic objects and slates. The user can also operate the system via built-in voice commands inputted into the embedded microphone. The system runs a special version of Windows 10 called Windows Holographic and supports

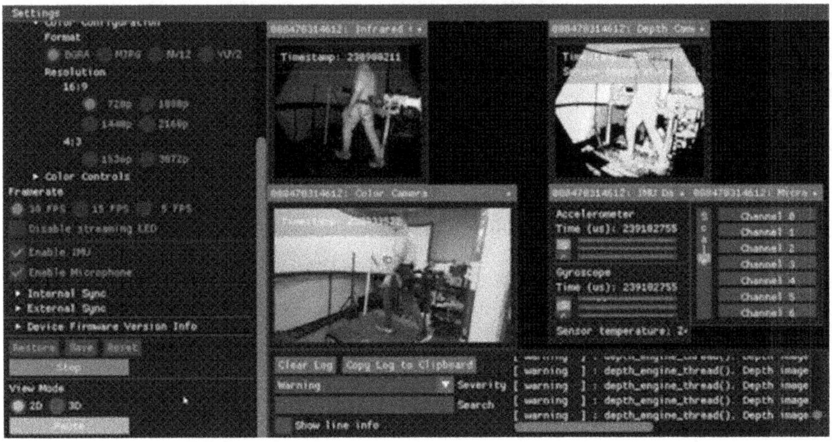

Figure 11.5 Azure Kinect DK User Interface.

a variety of applications, including those that have been designed from the ground up for the device and enhanced versions of traditional Windows applications such as Microsoft Teams. Special functionality known as spatial mapping scans the physical environment and enables users to anchor holographic objects to real-life surfaces. The device includes special audio to further immerse the user in the mixed reality environment. The headset includes a camera, embedded depth sensors, accelerometer, gyroscope, and magnetometer. Additional specifications can be found at www.microsoft.com/en-us/hololens/hardware#document-experiences.

Our system utilizes a specific application on the HoloLens titled Holographic Remoting, which connects to the host PC running the Unity chapter over a network, enabling Unity chapters with the MRTK3 plugin and the corresponding IP address to be deployed to the HoloLens.

Due to its seamless integration with the MRTK3, Azure Kinect DK, and all other hardware and software components in the Microsoft ecosystem, the Microsoft HoloLens 2 enabled the team to create a system that fulfilled the use cases of the chapter in a manner that no competing mixed reality headset could.

11.3.10 Alienware Host PC (x1) – Patient's Device

This device hosts the simulated setup that a patient would use at any given medical center. The host PC that connects via Ethernet to the treadmill and runs the Unity minigames that the patient plays is a prebuilt Alienware desktop computer that has been modified to include an Nvidia Titan V GPU in addition to the Nvidia 2080 Ti GPU. Nvidia designed the Titan V to include advanced machine learning capabilities that, with supporting technical expertise, would enable designers to create a three-dimensional avatar from a single camera's point cloud by procedurally generating the remaining portions of the model. The team ultimately decided to utilize three Azure Kinect DK cameras instead of the procedural generation method due to the superior accuracy and significantly decreased development time of meshing three point clouds into one cohesive model. The device also includes an Intel® Core™ i9-9820X CPU @ 3.30 GHz and 47.9 GB of RAM.

11.3.11 Dell Laptop (x1) – Doctor's Device

The system utilized a 17″ Dell laptop to run the 3D Capture and Communication software during demonstrations. The team utilized the laptop to simulate the doctor's computer to test voice communication via Unity multiplayer functionality over a network.

11.4 RESULTS

The chapter successfully completed the deliverables. Within the three-dimensional capture and communication subsystem, the Azure Kinect DK [13] cameras surround the patient and stream RGB-D data to the patient's PC. The cameras are positioned in a circular fashion equidistant from each other and angled in a manner that allows them to capture a 360-degree field of view of the patient while he or she operates the treadmill. The Unity chapter utilizes "Azure Kinect Examples for Unity," a collection of Unity assets that work with an open-source body-tracking algorithm, to convert each Kinect's RGB-D data into a point cloud in a Unity scene. The three point clouds have default x-coordinates, y-coordinates, and z-coordinates that merge to form a three-dimensional avatar within the Unity chapter. When configured for Holographic Remoting, entering play mode within Unity will deploy the Unity scene to the doctor's HoloLens 2. The patient's point cloud is updated in real-time. The doctor can manipulate the position, rotation, and scale of the patient's avatar to evaluate the physical and cognitive skills of the patient as he or she plays the game development subsystem's mixed reality minigames. The Photon Voice 2 plugin for Unity [14,15] enables voice communication between the doctor's HoloLens and patient's HoloLens over a network.

The mixed reality game development subsystem includes three minigames that the doctor can utilize to evaluate the patient's physical and cognitive skills post-stroke. The patient, who simultaneously operates a HoloLens and the treadmill, plays one of three mixed reality games designed to test his or her physical and cognitive abilities. The patient's actions determine the real-time movement that the 3D capture subsystem imports into the doctor's perspective. The team developed the games with the Unity game

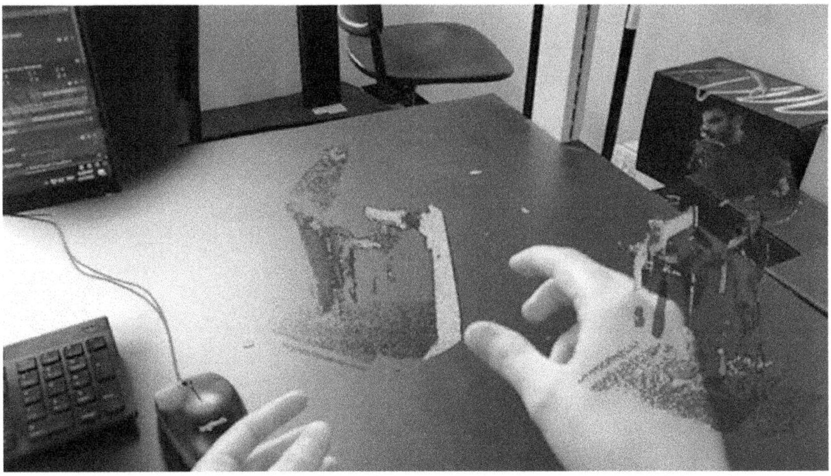

Figure 11.6 Unmerged point clouds within doctors' perspective.

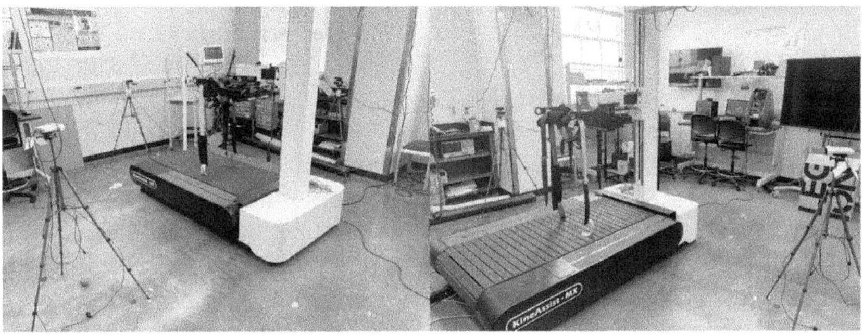

Figure 11.7 Integrated system for patient use.

engine, which contains an infrastructure built in C++ and enables developers to import custom C# scripts to enhance their chapters.

The following Figure 11.6 displays two of the three unmerged point clouds visualized within the doctor's HoloLens and the integrated system in which the patient will simultaneously operate the treadmill and HoloLens while the cameras create the model. Figure 11.7 shows a integrated system for patient use.

11.4.1 XR Game Development – Game 1: Speed Tracker

The first game aims to evaluate the patient's walking speed. The patient walks through a holographic environment depicting a nature trail within a forest biome. The treadmill tracks the patient's walking speed in meters

Figure 11.8 Game 1 demonstration.

per second. The Unity chapter imports the patient's walking speed from the treadmill and visualizes it on a holographic slate within the Unity scene. The game 1 demo is shown in Figure 11.8. The patient is faced by a series of rounds in which the walking speed required increases. Each round is ten minutes and must be completed faster than the threshold. The game provides an objective measurement for the recovery of the patient based on the number of rounds completed. The game takes place in an immersive forest environment. The Needed game object contains the script "Read Speed," which reads the treadmill speed input and places it on the slate.

The Canvas game object containing a Text Mesh Pro gameObject. This displays text and contains the scripts 'textPos' and 'text Update' which move the text in front of the player and provide continuous changing text. The Terrain object populates the mingames with trees and other natural elements to increase the immersion of the experience. The Goal object is an instance of GameObject represented as a white barrier indicating the threshold needed to pass the current round.

11.4.2 XR Game Development – Game 2: Obstacle Course

The second game aims to evaluate the patient's balance. The patient must walk forward and avoid incoming holographic objects by either stepping over or ducking under them. The components below allow this game to function as intended.

The StepOver/DuckUnder objects are GameObjects referred to in "HeadRef". Headref spawns these blocks at random intervals based on height. The Needed game object containing the script "Read Speed," which

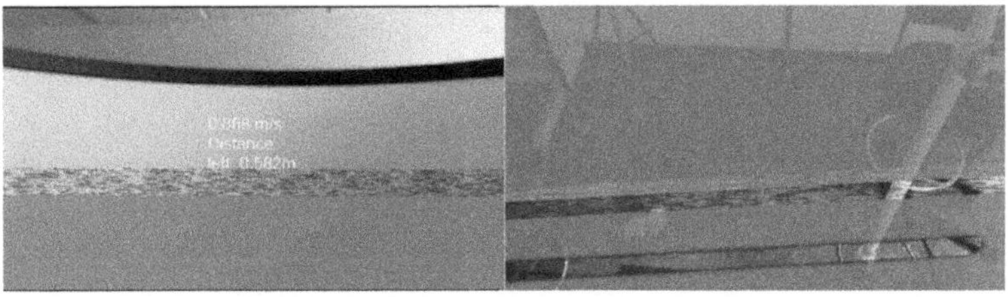

Figure 11.9 Game 2 demonstration.

reads the treadmill speed input and places it on the slate. The Ground Terrain object includes the HDRP terrain used for applying textures, trees, grass, and mountains. The HeadRef object populates the game with trees and other natural elements for an immersive experience. To learn more about the Audio + PointCloudServer object, refer to the 3D capture and communication documentation.

11.4.3 XR Game Development – Game 3: Grocery Store Simulator

The third game evaluates the patient's ability to decipher and categorize objects by simulating a trip to the grocery store. The patient must walk down a grocery isle, observe the requested contents of a grocery list displayed on a holographic slate, grab the corresponding items off the shelves, and place them in a grocery basket based on whether the item is a fruit or a vegetable. Food spawns randomly, and the game continues until the patient completes the list.

The Needed game object containing the script "Read Speed," which reads the treadmill speed input and places it on the slate. The TriggerEnter game object contains the script "Read Speed," which reads the treadmill speed input and places it on the slate. A script created to read a GameObjects properties. If the properties are that of fruit, allow the fruit to be crossed off the list. If properties are !fruit, the patient needs to pick another object. This script acts as a "sorting basket." You can make this region invisible so one can place it in a basket. The FruitMaster script controls the spawning of the fruit. The fruit will spawn randomly along the shelves' position. For more information on the Audio + PointCloudServer object, refer to 3D capture and communication documentation.

11.5 CONCLUSION

The chapter successfully enables doctors and patients to utilize the system to complete stroke rehabilitation medical appointments in augmented reality from any physical location. The chapter will eliminate the transportation

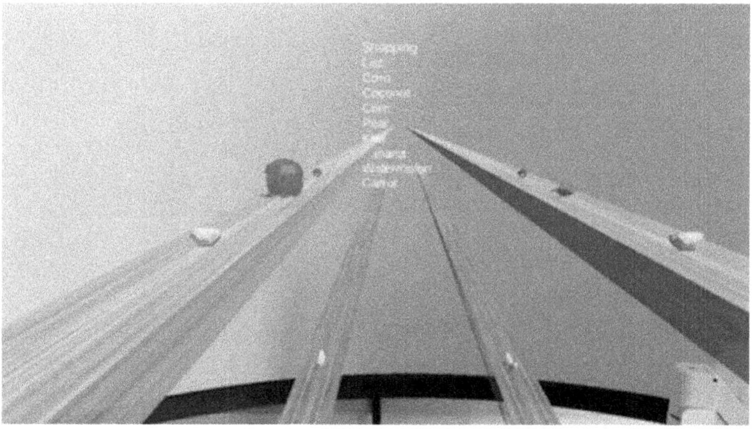

Figure 11.10 Game 3 demonstration.

requirement for doctor's appointments, decreasing the amount of time required for a post-stroke patient to receive therapeutic treatment. Future iterations of the chapter could improve the visual clarity of the patient's avatar by replacing the standard point clouds with a point cloud mesh and increasing the framerate beyond the 15 fps minimum threshold stated in the chapter deliverables. The chapter yielded valuable evidence demonstrating the impact of utilizing augmented reality in the medical field.

11.6 RELATED WORKS

Literary works based on the research of various groups elaborate on the realized and unrealized potential of integrating mixed reality into the medical industry. Alamri et al. (2008) developed an augmented reality rehabilitation system called AR-REHAB, which was shown to be effective in improving motor skills in post-stroke patients. The system also enables progress tracking over time and offers engaging exercises designed to motivate patients to complete their treatments.The paper denotes the significant impact of strokes, one of the world's leading causes of death, on individual patient health and society.

Burke et al. [17] further support the effectiveness of emerging mixed reality technologies for providing engaging stroke rehabilitation exercises with a focus on the upper body. The paper details the development of mixed reality with an emphasis on eye-tracking and hand-tracking technologies, which enable developers to create exercises that test the cognitive skills and motor skills of the upper body. Our system's exercises, which test patients' walking speed, balance, and ability to duck under objects and step over objects, have a special emphasis on the lower body; the paper therefore provides valuable information that further elaborates on the effectiveness of such systems on areas that represent a secondary focus of our system.

Cho et al. (2023) explains the potential impact and limitations of the Metaverse on stroke rehabilitation. Coordinating the Metaverse with a virtual reality-assisted treadmill benefits patients' cognition, motor skills, muscular endurance, respiratory and cardiovascular function, etc. However, creating an infrastructure with the emerging and expensive technology could limit the potential widespread use of the technology in the medical field.

The influence of mixed reality will not solely be contained within the medical industry. Sameer Mosa AlNajdi and the Education Technology Department at the University of Tabuk (2022) discuss the impact of mixed reality on education. The article discusses how the Saudi Ministry of Education embedded augmented reality technology into QR codes in textbooks to enhance student learning during the COVID-19 pandemic. The students who utilized the technology performed more effectively than those who did not.

ACKNOWLEDGMENT

Nashmi Alheraijy, Qais Altannak, Trent Boughner, Patrick Brooks, Luke Burbee, and Zane Pautzke thank Dr. Fei Hu, *Professor and Faculty Advisor*, Dr. Sushma Kotru, *Professor*, Shengting Cao, *Advisor*, Jennifer Hinton, *Purchasing Assistant*, and The University of Alabama Electrical and Computer Engineering (ECE) Department.

REFERENCES

1. Alamri, A., J. Cha, and A. El Saddik. "AR-REHAB: An augmented reality framework for poststroke-patient rehabilitation," *IEEE Transactions on Instrumentation and Measurement*, vol. 59, no. 10, pp. 2554–2563, Oct. 2010. https://doi.org/10.1109/TIM.2010.2057750.
2. AlNajdi, S. M. "The effectiveness of using augmented reality (AR) to enhance student performance: Using quick response (QR) codes in student textbooks in the Saudi education system," *Educational Technology Research and Development*, vol. 70, no. 3, pp. 1105–1124, 2022. https://doi.org/10.1007/s11423-022-10100-4. Epub 2022 Apr 4.
3. "Azure-Kinect-Sensor-SDK/docs/usage.md at develop microsoft/Azure-Kinect-Sensor-SDK," *GitHub*, 2024. https://github.com/microsoft/Azure-Kinect-Sensor-SDK/blob/develop/docs/usage.md. Accessed: Apr. 29, 2024.
4. "Azure Kinect and Femto Bolt Examples for Unity | Integration | Unity Asset Store," *assetstore.unity.com*, 2024. https://assetstore.unity.com/packages/tools/integration/azure-kinect-and-femto-bolt-examples-for-unity-149700. Accessed: Apr. 29, 2024.

5. joyjaz, "Microsoft HoloLens," *Microsoft Learn*, 2024. https://learn.microsoft.com/en-us/hololens/.
6. Cho, K.-H., J.-B. Park, and A. Kang, "Metaverse for exercise rehabilitation: Possibilities and limitations," *International Journal of Environmental Research and Public Health*, vol. 20, no. 8, p. 5483, Jan. 2023. https://doi.org/10.3390/ijerph20085483.
7. Northwestern Medicine, "KineAssist-MXTM," *Northwestern Medicine.* https://www.nm.org/conditions-and-care-areas/treatments/kineasst-mx.
8. "Photon Voice 2 | Audio | Unity Asset Store," *assetstore.unity.com.* https://assetstore.unity.com/packages/tools/audio/photon-voice-2-130518
9. "Azure Kinect Body Tracking SDK download," *learn.microsoft.com*, Sep. 02, 2022. https://learn.microsoft.com/en-us/azure/kinect-dk/body-sdk-download
10. "Azure Kinect DK documentation," *learn.microsoft.com.* https://learn.microsoft.com/en-us/azure/Kinect-dk/
11. satyapoojasama, "Add the mixed reality OpenXR plugin to your unity chapter – mixed reality," *learn.microsoft.com*, Nov. 22, 2022. https://learn.microsoft.com/en-us/windows/mixed-reality/develop/unity/mixed-reality-openxr-plugin
12. Sean-Kerawala, "Welcome to the mixed reality feature tool – mixed reality," *learn.microsoft.com*, Dec. 14, 2022. https://learn.microsoft.com/en-us/windows/mixed-reality/develop/unity/welcome-to-mr-feature-tool
13. tesych, "Synchronize multiple Azure Kinect DK devices," *learn.microsoft.com*, Sep. 21, 2022. https://learn.microsoft.com/en-us/azure/kinect-dk/multi-camera-sync (accessed Apr. 29, 2024).
14. Unity Technologies, "Create high-quality graphics and stunning visuals | Unity HDRP," *unity.com.* https://unity.com/srp/High-Definition-Render-Pipeline#:~:text=Access%20cutting%2Dedge%20real%2Dtime
15. Unity Technologies, "Get started with unity – download the unity hub & install the editor," *unity.com.* https://unity.com/download
16. Unity Technologies, "Unity 2021.3.7," *unity.com*, Jul. 28, 2022. https://unity.com/releases/editor/whats-new/2021.3.7 (accessed Apr. 29, 2024).
17. Burke, J. W., M. D. J. McNeill, D. K. Charles, P. J. Morrow, J. H. Crosbie, and S. M. McDonough. "Augmented reality games for upper-limb stroke rehabilitation," in *Proceedings of the 2010 Second International Conference on Games and Virtual Worlds for Serious Applications*, Braga, Portugal, 2010, pp. 75–78. https://doi.org/10.1109/VS-GAMES.2010.21.

Chapter 12

The Metaverse for Medical Rehabilitation Applications

Ruxin Zheng, Lifan Xu, and Shunqiao Sun

12.1 INTRODUCTION

Exercise rehabilitation serves as a pivotal approach to restoring physical, mental, and social well-being, particularly for the elderly and those facing physical challenges. This holistic health service not only aims to assist individuals in performing daily life activities but also acts as a comprehensive prevention and recovery program for overall health. However, the demand for exercise rehabilitation programs, especially for the disabled, elderly, and patients, significantly exceeds the current supply [1, 2]. This insufficiency is exacerbated by the challenges posed by traditional rehabilitation methods, predominantly manual techniques, and resource-intensive processes within rehabilitation centers [3, 4].

The global landscape of healthcare faces additional complexities, particularly for individuals requiring personal care assistance, as routine medical care becomes challenging during pandemic-related restrictions. The heightened infection risk and social distancing measures create continuous background stress for people with disabilities, impacting their ability to access essential personal care assistance services [5, 6].

Moreover, the traditional training content in rehabilitation often involves repetitive and mechanical movements, leading to patient boredom and reduced effectiveness of rehabilitation, especially in pediatric cases [7]. In response to these challenges, there is a pressing need for innovative approaches that not only address the shortcomings of conventional rehabilitation but also cater to the diverse needs of the population [8].

Recent developments in the field of exercise rehabilitation have embraced virtual solutions to enhance engagement and effectiveness. Virtual exercise rehabilitation programs, specifically designed for weak and injured populations, aim to address issues of monotony and lack of motivation. One promising avenue in this transformative journey is the Metaverse, an interactive simulation that enables users to experience virtual spaces akin to reality. The metaverse leverages advanced technologies, such as virtual reality (VR), augmented reality (AR), mixed reality (MR), interactable robotics, and wearable devices to provide immersive and interactive experiences.

DOI: 10.1201/9781032663609-15

This chapter aims to provide a comprehensive exploration of the diverse applications of metaverse technologies in the realm of rehabilitation. Specifically, we delve into the integral components of the metaverse, namely VR, AR, and MR, elucidating their transformative potential in the rehabilitation landscape. Additionally, we examine the critical aspect of human-computer interaction, encompassing the incorporation of robotics and wearable devices. By doing so, we aim to present a nuanced understanding of how these interconnected elements within the metaverse contribute to the evolution of rehabilitation practices.

12.2 VR-BASED REHABILITATION

Immersive technologies such as VR, AR, and MR have ushered in a new era in rehabilitation. VR transports users to a fully immersive digital world, using head-mounted displays and data gloves to minimize external distractions. AR enhances the physical world with digital overlays through devices such as smartphones or tablets. MR blends real and virtual worlds, allowing interaction with both, while XR encompasses all these realities and future technologies [9, 10]. This section will explore their transformative roles in enhancing patient engagement and the effectiveness of rehabilitation therapies.

12.3 ADVANTAGES OF VR-BASED REHABILITATION

Traditional rehabilitation offers the advantage of proven long-term benefits through personalized, hands-on therapy from skilled professionals. While traditional rehabilitation has demonstrated long-term benefits, its reliance on subjective therapist assessments and the necessity for extensive, one-on-one sessions can lead to variable outcomes and significant costs [11]. The repetitive nature of treatment exercises may also diminish patient motivation over time due to their monotonous nature [12]. Additionally, the pandemic has introduced new challenges for traditional rehabilitation; in-person sessions carry an increased risk of infection, and social distancing measures impose additional stress on individuals with disabilities, further complicating their treatment and recovery process [5, 6].

In response to these challenges, innovative technologies such as VR, AR, and MR have emerged, transforming the rehabilitation landscape [13]. These technologies offer substantial benefits, motivating patients for long-term treatment and standardizing care quality for stroke survivors [14, 15]. VR creates realistic, human-like virtual environments using computer technology, which has been evolving rapidly to provide more lifelike settings for recovery induction. AR, through smart devices, blends the real and virtual worlds, enabling the exploration of concepts and phenomena not accessible in traditional clinical settings. It's particularly recognized for its effectiveness in stroke rehabilitation due to its capacity for intense, repetitive, and context-specific exercises. MR integrates the physical and digital worlds, allowing real-time interaction between physical entities and digital elements, offering an engaging medium for performance communication and self-assessment. Figure 12.1 illustrates the real to virtual difference between AR, VR, and MR.

VR-based rehabilitation stands as a compelling alternative to traditional methods, blending immersible multimedia technologies with rehabilitative practices to create interactive and engaging environments. This approach, which includes visual, auditory, and tactile elements, greatly enhances patient engagement and comfort, allowing therapy to be conducted in the patient's home [11, 16, 17]. VR-type technology provides detailed feedback and allows for real-time progress tracking, addressing the subjective limitations of conventional therapy. Its efficacy in treating motor impairments associated with diseases like Parkinson's, multiple sclerosis, cerebral palsy, and stroke has led to better outcomes than traditional therapy [12, 18–23].

Innovative VR-based rehabilitation applications often utilize a range of technologies, such as head-mounted displays (HMDs) and hand controllers, to develop exercises that are not only purposeful but also customizable to meet the unique needs and abilities of each patient. The exploration of VR-type technology in rehabilitation began as early as 2004, with key studies highlighting its practicality and effectiveness. A pioneering example is the VR training environment created by Jaffe et al. [24], which combined a motorized treadmill, HMD, and specially designed shoes with switches to help patients practice stepping over virtual obstacles.

Since then, a wide array of VR-based rehabilitation systems have been developed, yielding promising results across various applications. These systems cover a broad spectrum of rehabilitation needs, ranging from the upper extremities (including arms, hands, fingers, and shoulders) to the lower extremities (focusing on ankles, gait, balance, and posture) [6]. Table 12.1 provides a comprehensive overview of the representative VR-based rehabilitation systems developed in recent years, illustrating the broad

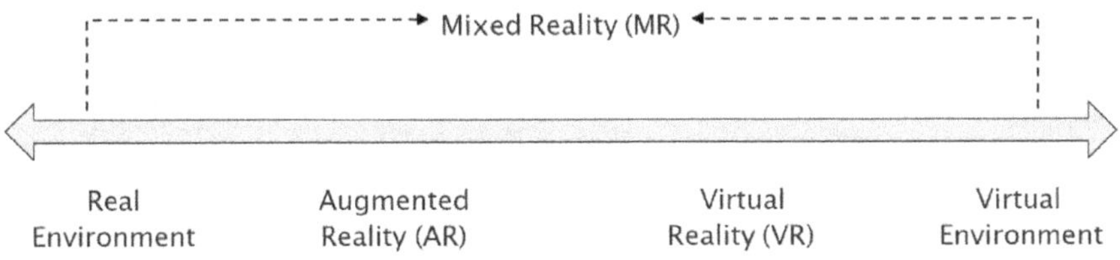

Figure 12.1 Continuum of Real to Virtual Environments.

This diagram illustrates the spectrum of environments from real to virtual, highlighting the transition from the physical world (Real Environment) through augmented reality (AR), to the fully immersive experience of VR, and the intermediate phase of mixed reality (MR).

Table 12.1 VR-based rehabilitation studies/systems in recent years. Aberrations: electromyography (EMG)

Article	*Setting*		*Goals*	*Devices*
	Lab	*Hospital*		
[25]	√		Arms, Wrists, Shoulders, Elbows, Muscle Strength, Range of Motion	EMG, RGB Camera, Fiducial Markers
[26]	√		Shoulders, Muscle Strength, Range of Motion	EMG, RGB Camera, Fiducial Markers
[27]	√		Arms, Hands/Fingers, Shoulders, Elbows, Balance, Postural, ADL	RGB Camera, Fiducial Markers
[28]	√		Gait, Balance, Postural	HMD, RGB Camera
[29]	√		Arms, Hands/Fingers, Wrists	HMD
[30]	√		Hands/Fingers, ADL	RGB Camera
[31]	√		Arms, Hands/Fingers, Wrists, Shoulders, Elbows	Kinect
[32]	√		Hands/Fingers, ADL	Table Projection, Depth Camera
[33]	√		Shoulders, Elbows	Kinect
[34]		√	Arms, Hands/Fingers, Wrists, Shoulders, Elbows, ADL	Table Projection, Kinect
[35]		√	Hands/Fingers	RGB Camera, Fiducial Markers
[36]	√		Hands/Fingers, Wrists, Elbows	RGB Camera
[37]	√		Hands/Fingers	RGB Camera
[38]	√		Shoulders, Elbows, Range of Motion	Kinect
[39]	√		Ankles, Muscle Strength, Gait, Balance	RGB Camera
[40]	√		Arms	Kinect, RGB Camera
[41]		√	Arms, Hands/Fingers, Wrists, Shoulders, Elbows, Range of Motion	Kinect
[42]	√		Arms, Shoulders, Elbows	Kinect

applicability and flexibility of VR technology. This diversity in applications underscores the versatility and adaptability of VR technology in addressing different rehabilitation challenges, offering a tailored approach to patient recovery and therapy.

12.4 COLLABORATIVE VIRTUAL ENVIRONMENTS

Collaborative virtual environments (CVEs) represent a significant breakthrough in re- habilitation practices, offering a unique blend of technology and human interaction. These advanced platforms create shared, interactive spaces in simulated worlds, leveraging the capabilities of VR, AR, and MR to facilitate real-time collaboration and communication. This innovative approach bridges geographical gaps and adds a crucial social dimension to the rehabilitation process [22].

The human need for interpersonal relationships and social interaction is a core aspect of CVEs in rehabilitation. Theories such as Maslow's hierarchy of needs and Ryan and Deci's concept of relatedness highlight the importance of social connections in human motivation [43]. For populations such as older adults and individuals with disabilities, including stroke survivors, the risk of social isolation and loneliness is high, potentially leading to severe health issues such as cognitive decline and Alzheimer's disease [19]. CVEs address these challenges by enabling patients, therapists, and family members to interact in a virtual space, thus enhancing motivation, engagement, and overall therapy effectiveness.

Social facilitation, or the impact of others' presence on performance, plays a crucial role in the efficacy of CVEs in rehabilitation. Studies have shown that social interactions, even with virtual avatars, can significantly boost motivation and effort during rehabilitation [44, 45]. Stroke patients, in particular, benefit from CVEs as they adapt to changes in daily activities and social roles, especially when they return home early due to advancements in rehabilitation techniques and the need to reduce public health costs. CVEs support home exercises, addressing the common challenge of lacking motivation and supervision [46].

Web-based CVEs present significant advantages that bolster their utility and appeal in VR applications. These environments adeptly simulate training scenarios that might be logistically challenging or physically demanding in the real world. Leveraging cross-browser graphical libraries, they render intricate virtual spaces directly through user web browsers, facilitating a seamless and accessible user experience.

A prime example of this technology is the LIRKIS G-CVE (Global Collaborative Virtual Environment), which exemplifies a web-based VR/MR system accessible globally over the internet [47]. The architecture of LIRKIS G-CVE, as illustrated in Figure 12.2, encompasses a remote web server and a web-client interface, along with additional modules and interfaces for enhanced functionality.

The key strengths of web-based CVEs include the swift development of communication interfaces and cross-platform compatibility, supporting a wide range of devices and operating systems. Their design prioritizes global accessibility, ensuring that users can connect and collaborate irrespective

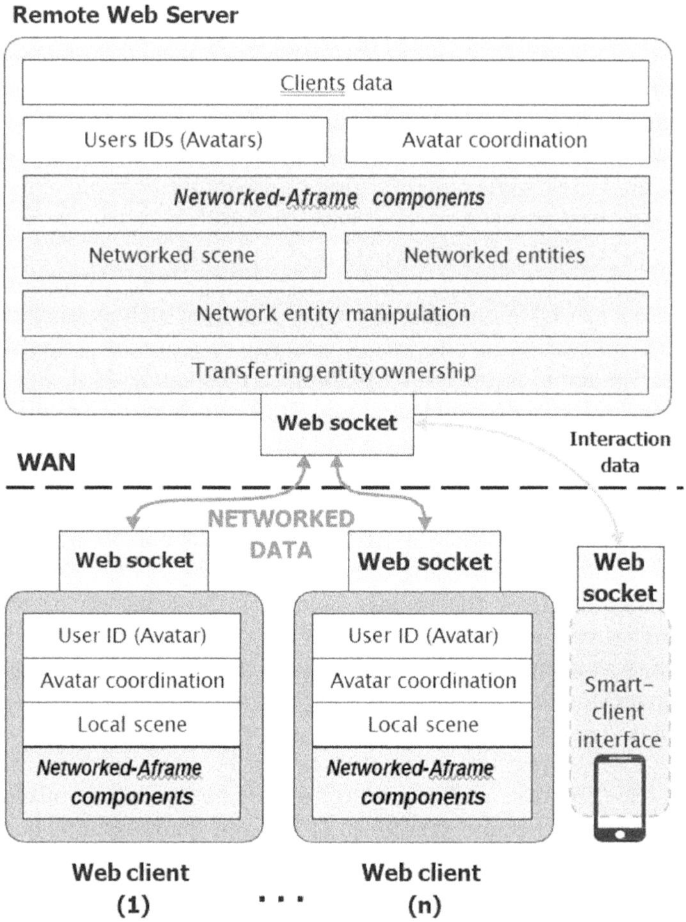

Figure 12.2 The web-based LIRKIS G-CVE system model.

of their physical locations. Additionally, these systems offer a cost-effective alternative to standard VR applications, making them a more viable option for widespread implementation [22]. These attributes collectively render web-based CVEs not just a practical option for rehabilitation purposes but also an inclusive and versatile solution catering to the diverse needs of users worldwide.

12.5 ADAPTING GAMING TECHNOLOGIES FOR REHABILITATION

Interactive video games and VR technology have significantly evolved, branching into categories for entertainment, learning, and behavioral change. The integration of these games in rehabilitation began with devices like Sony PlayStation II's EyeToy Camera and Nintendo Wii. These technologies, combining motion capture with gaming, have proven effective in both home and clinical rehabilitation settings, fostering the development of applications involving robotics, VR, and tangible interfaces. Adapting these entertainment-focused systems for rehabilitation, however, presents challenges such as designing suitable controllers for individuals with limited hand strength and tailoring game mechanics to meet therapeutic goals. Moreover, games must be adaptable for varying severities of conditions such as hemiparesis and accessible to users with cognitive impairments. Despite these challenges, technological advancements have made VR systems such as Oculus Rift, PlayStation VR, HTC Vive, and Microsoft Hololens more affordable and accessible, expanding their application in diverse therapeutic contexts [12].

The Rehabilitation Gaming System (RGS) is a VR-based tool designed for diverse rehabilitation needs. Utilizing a unique action execution and observation paradigm (see Figure 12.3), it enables users to control a virtual avatar, promoting movement coordination. This approach is based on the theory that action observation activates motor areas, such as mirror neurons, promoting neuronal plasticity and functional recovery from various rehabilitation needs. It features adaptive elements, adjusting task difficulty and movement trajectory to optimize user motivation and learning. By amplifying virtual movements to match intended actions, RGS accommodates patients' physical limitations, ensuring engagement and progress in rehabilitation [19].

In [12], "Project Star Catcher" is introduced as a VR-based rehabilitation game designed to assist individuals with hemiparesis. This game adapts

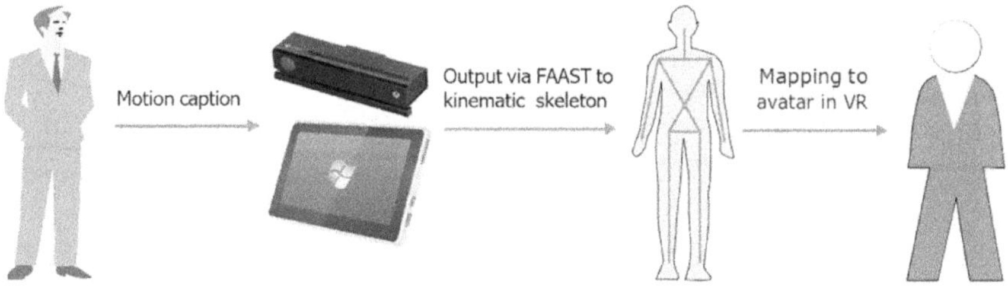

Figure 12.3 RGS system configuration and operation: User movements are captured by the Kinect motion capture device, with actions translated onto the avatar's virtual limbs using the Flexible Action and Articulated Skeleton Toolkit (FAAST).

the principles of modified Constraint Induced Therapy (mCIT) into a VR setting. Instead of physically restricting a patient's stronger arm, the game uses a virtual incentive system to encourage the use of the weaker arm, capitalizing on the immersive capabilities of the HTC Vive. This innovative approach not only engages users but also ensures they adhere to the movement protocols integral to mCIT, enhancing the overall rehabilitation experience.

Similarly, [20] introduces "PhyRe Up!," a telerehabilitation system employing MR and gamification to aid stroke patients in home-based physical rehabilitation. This system utilizes MR technology, including the Microsoft HoloLens 2 and Azure Kinect DK, to provide real-time feedback and instructional guidance, thereby enhancing exercise precision and supporting patient autonomy. The combination of virtual and physical elements within "PhyRe Up!" creates an immersive environment that simulates a genuine rehabilitation session, optimizing the therapist-patient interaction for improved exercise customization and assessment.

The importance of social interaction in rehabilitation, as discussed in Section II-B, is further exemplified in [43], where immersive VR biking is used to encourage social interaction in rehabilitation. Here, users participate in a tandem biking challenge set in a virtual mountainous landscape, with the game's design catering to various physical abilities. Additionally, [19] explores how adaptive mapping in RGS can be used for multiplayer interaction between patients and family members, enhancing social engagement. They developed an air hockey-like game that emphasizes speed and precision, crucial elements in rehabilitation tasks, to promote social interaction and acceptance.

12.6 CHALLENGES OF VR-BASED REHABILITATION

VR-based rehabilitation brings numerous advantages to the table, offering innovative and immersive ways to enhance patient care. However, its integration into rehabilitation practices is faced with significant challenges, encompassing financial, technical, psychological, and physiological aspects [48]. These issues need careful consideration to ensure the effective and safe application of VR in therapeutic settings.

One of the foremost barriers is the cost and accessibility of VR technology. While hardware such as HMDs is becoming more affordable, the software for creating personalized rehabilitation programs remains prohibitively expensive. This is a particular challenge for individual therapists who do not have the financial support of larger healthcare institutions. Compounding this issue is the lack of comprehensive research on the cost-effectiveness of VR and AR in healthcare, making it difficult for therapists to justify the investment in these technologies.

Another significant challenge is the lack of standardization and research in the field. The literature is notably sparse on studies evaluating the adoption rate of VR systems among patients, leaving therapists without a clear understanding of how widely these technologies might be accepted and used in their practices. The absence of controlled trials or systematic reviews means that the integration of VR into rehabilitation is often based on limited empirical evidence, rather than robust clinical data.

The psychological and emotional impact of VR on patients is also a major concern. The immersive nature of VR can induce intense emotional responses, especially when it involves reliving fears or traumatic events. This heightened emotional experience can lead to psychological distress and challenges in differentiating VR experiences from real-life events, particularly for patients with existing mental health conditions or those at risk of deterioration.

Physical and psychological side effects present another set of challenges. Common issues such as motion sickness, nausea, dizziness, and vomiting can significantly impact a patient's ability to engage in daily activities and can be exacerbated by frequent or prolonged VR sessions. The potential of VR to alter body image perception is another concern, as it could lead to negative self-perception and body dysmorphia, especially if the VR simulations are not carefully designed.

The cognitive and neural overload caused by VR is yet another hurdle. The immersive and stimulating environment can lead to an overload of neural and cognitive resources, resulting in increased stress levels and a significant "wear and tear" response. Moreover, the experience of embodying a virtual avatar different from one's physical self can lead to changes in cortical excitability and reorganization, potentially affecting the patient's real-life motor skills and spatial awareness.

Finally, the integration of VR technology into existing rehabilitation protocols poses its own set of challenges. The lack of standardized guidelines for the use of VR in therapy leads to inconsistencies in treatment methodologies and difficulties in measuring the effectiveness and safety of VR-based interventions.

In summary, the challenges facing VR-based rehabilitation are diverse and complex, ranging from financial and technical issues to psychological and physiological impacts on patients. These challenges must be addressed comprehensively to harness the full potential of VR in therapeutic contexts.

12.7 REHABILITATION ROBOTICS

Rehabilitation robotics is an emerging field that combines advanced robotic technology with therapeutic techniques to improve rehabilitation for patients with a variety of motor function impairments [49]. This area focuses on designing robotic devices tailored to assist in the rehabilitation of specific body parts, such as arms, hands, legs, and ankles, and on innovating therapeutic training and assessment methods. Gaining acceptance among patients, rehabilitation robotics has shown to be a valuable addition to conventional therapies, particularly for those recovering from strokes. Situated within the realms of biomedical engineering and human-robot interaction, rehabilitation robotics is a collaborative effort involving clinicians, therapists, and engineers. Key goals of this field are to develop technologies that are user-friendly for both patients and healthcare professionals, enhance the effectiveness of standard therapies, and improve the daily lives of patients. Robotic systems in rehabilitation are capable of providing more intense, consistent, and extended therapy sessions compared to traditional physical therapy, thereby advancing the recovery process for patients with neurological issues and providing therapists with better tools for tracking and assessing patient progress [7].

Having established the foundational understanding of rehabilitation robotics and its role in improving motor function impairments, let us delve deeper into its specific aspects. The following subsections will explore the "Advantages of Rehabilitation Robotics," "Wearable Robotic Technology," and "Socially Assistive Robotic," each offering unique insights into how this field is revolutionizing patient care and therapy practices.

12.8 ADVANTAGES OF REHABILITATION ROBOTICS

Rehabilitation robotics offers a range of key benefits that significantly enhance the quality and effectiveness of therapeutic processes [50]. One of the foremost advantages of these robotic systems is their ability to provide precise and consistent therapeutic interventions. With the integration of advanced sensors, rehabilitation robots can meticulously record detailed movement data, including position, velocity, and joint torques. This capability ensures accuracy in therapeutic exercises, thereby enhancing the rehabilitation process's effectiveness.

Another significant advantage of robotic rehabilitation is the delivery of high-dosage and high-intensity training. Robots can consistently administer a large number of repetitive movements in a single session, a level of intensity often required for significant improvements in motor control [51]. This is particularly challenging to achieve through traditional physical therapy methods. Moreover, these robotic systems enable objective measurement and progress tracking, assessing a patient's initial impairments and monitoring changes in response to treatment. This precision in data collection minimizes the impact of interrater variability on outcome assessments and facilitates complex biomechanical analyses, allowing therapists to make more informed decisions regarding treatment plans [52].

Additionally, rehabilitation robots play a crucial role in reducing the physical and cognitive burden on therapists. By automating the repetitive and labor-intensive aspects of treatment, therapists can focus on more complex patient care aspects. This not only reduces the strain on therapists but also enhances the overall quality of patient care.

Task-oriented training is another area where rehabilitation robotics excels [53]. These systems can assist patients in performing specific, prescribed movements with greater precision and repeatability than is possible with manual therapy. This method has shown promising results, often surpassing conventional exercises focused on passive movement of impaired joints.

Extending beyond the clinical settings, robotic technology has also made its way into home-based rehabilitation [54]. Patients can now perform prescribed exercises with robotic assistance at home, reducing the need for frequent visits to treatment centers. This approach not only facilitates intensive therapy but also supports in-person or virtual therapy sessions. It has the potential to reduce costs and motivate patients to engage more actively in their therapeutic activities.

In conclusion, rehabilitation robotics brings transformative advantages to therapy, ranging from enhancing precision and consistency in exercises to enabling intensive training and objective progress tracking. By alleviating the burden on therapists and facilitating targeted, task-oriented training, these robotic systems are reshaping the landscape of rehabilitation, offering more effective and accessible treatment options for patients with diverse motor impairments.

12.9 WEARABLE ROBOTIC TECHNOLOGY

The field of robotic rehabilitation has predominantly focused on techniques that involve physical contact, often described as hands-on or wearable robotic technology. This approach has revolutionized rehabilitation, transforming it into a process where patients engage in guided, passive exercises facilitated by robots. These robots assist patients in correctly exercising the affected limb, ensuring proper execution and safety.

Rehabilitation robots involving physical contact are broadly categorized into end-effector and exoskeleton types, each with unique working principles as highlighted in [55]. End-effector robots interact at the limb's distal end, such as the hand in upper-limb rehabilitation, indirectly influencing the movement of the entire limb. An example of end-effector robotics is the end-effector-based upper-limb rehabilitation robot (EULRR), detailed in [56]. This robot features an Assist-as-Needed (AAN) controller, facilitating active rehabilitation training by allowing patients to move their arms within a virtual channel along a set path. If the patient's arm strays, the EULRR gently guides it back, ensuring exercises are performed correctly. Unlike traditional planar end-effectors, the EULRR allows for three-dimensional movement, enhancing the scope and effectiveness of rehabilitation.

In contrast, exoskeleton robots are wearable devices designed to mimic human anatomy closely, enabling targeted and precise rehabilitation exercises. This design is especially beneficial for addressing specific motor impairments, offering specialized and accurate exercises. A key advantage of exoskeleton robots is their adaptability; they can dynamically assess and respond during the rehabilitation process. This adaptability transforms exoskeletons from general rehabilitation tools into sophisticated therapeutic devices for targeted interventions. Various exoskeleton robots have been developed for both upper- and lower-extremity rehabilitation, demonstrating their versatility and efficacy in diverse rehabilitation scenarios, as noted in [55, 57]. Several market-leading examples showcase the advancements in rehabilitative exoskeleton technology. Harmony, for instance, is an advanced upper-extremity rehabilitation exoskeleton. This system features a sophisticated five-degree-of-freedom shoulder assembly, coupled with mechanisms for single-degree-of-freedom motion at the elbow and wrist. LOKOMAT stands out in aiding stroke survivors with automated treadmill training, enhancing their gait rehabilitation. In the realm of lower-limb assistance, REX offers a self-balancing exoskeleton that users can control via a joystick, promoting independent movement. Similarly, ReWalk integrates crutches and sensors to support patients who retain some mobility and balance capabilities. Illustrations of these innovative upper- and lower-limb exoskeletons are presented in Figure 12.4, highlighting their design and functional diversity.

Despite their advancements, current exoskeletons face challenges such as limited adaptability, delayed responses, continuous control, high costs, bulkiness, and extensive training requirements. Addressing these issues, the research in [58] introduces an artificial intelligence (AI)-powered Smart and Lightweight Exoskeleton System (AI-IoT-SES). This system integrates data from various sensors with AI and Internet of Things (IoT) technologies to intelligently classify information and generate precise rehabilitation commands. Specifically designed to enhance support and rehabilitation for paralyzed patients in smart, connected communities, the AI-IoT-SES significantly improves the assistance provided by caretakers. This innovative development represents a considerable advancement in exoskeleton technology, making these devices more adaptable, responsive, user-friendly, and accessible across various rehabilitation settings.

Furthermore, the emergence of wearable haptic devices presents significant opportunities in VR-based rehabilitation. These devices offer cutaneous feedback and are inherently suitable for immersive VR and neurorehabilitation scenarios. Their wearability ensures ease of use and the preservation of user dexterity, which is crucial for providing feedback during manipulative tasks. Various designs focus on distinct aspects of haptic

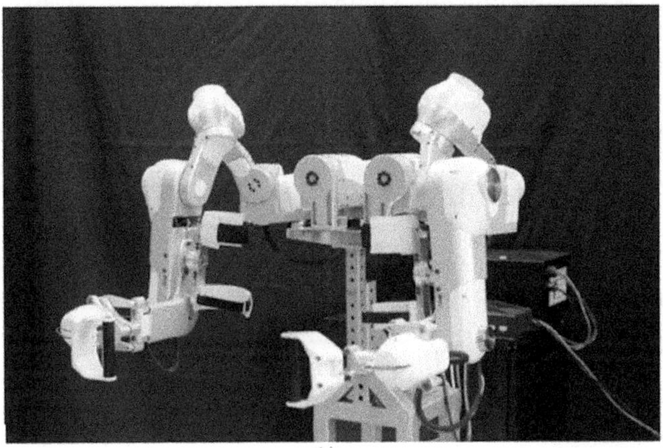

(a)

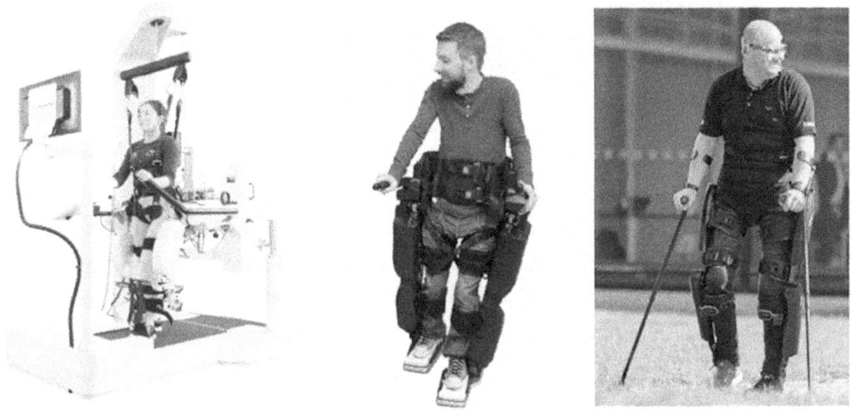

(b) (c) (d)

Figure 12.4 Exoskeleton robots for rehabilitation: (a) Harmony rehabilitation exoskeleton robot, (b) LOKOMAT, (c) REX, (d) ReWalk.

rendering, such as displaying contact transitions, surface orientation, and force feedback through skin deformation of the fingerpad. Devices following design guidelines in [59, 60] employ parallel kinematics wrapped around the finger, resulting in compact, lightweight devices that minimally interfere with finger movements. These advancements have improved user manipulation capabilities in VR and provided additional information about the physical characteristics of virtual objects [61].

12.10 Socially Assistive Robotic

Wearable robotic technology in rehabilitation, commonly known as assistive robotics, involves direct physical interaction between the patient and the robot, posing significant challenges, especially in terms of patient safety. Addressing these challenges is a crucial area of research. However, the requirements for safety compliance can significantly increase costs for both patients and healthcare providers, making such systems less accessible. Additionally, assistive robotics often fall short of promoting active patient participation. They also present integration difficulties in occupational therapy, which is primarily centered on functional activities, as discussed in [62].

Clinical research has shown that patient motivation is crucial for the success of neural rehabilitation therapies, with patient engagement being a direct influence on treatment outcomes. As shown in Figure 12.5, this new area has emerged from the combination of assistive robotics and social interactive robotics. A growing research trend is to incorporate motivational factors, including social interaction, into robotic-based rehabilitation. This approach has led to the development of Socially Assistive Robotics (SAR), which blends assistive robotics with social interactive capabilities [7]. SAR represents a paradigm shift in robotic rehabilitation, moving from solely physical contact-based assistance to including hands-off robotic support. This change reduces safety risks and eases the integration of these systems into clinical settings. SAR platforms are designed to enhance patient involvement and motivation in treatment, recognizing that repeated exercise is beneficial for recovery. The field of SAR is seen as one of the most challenging yet promising areas in contemporary rehabilitation robotics, aiming to bring a more holistic approach to patient care by merging technological assistance with elements of human interaction.

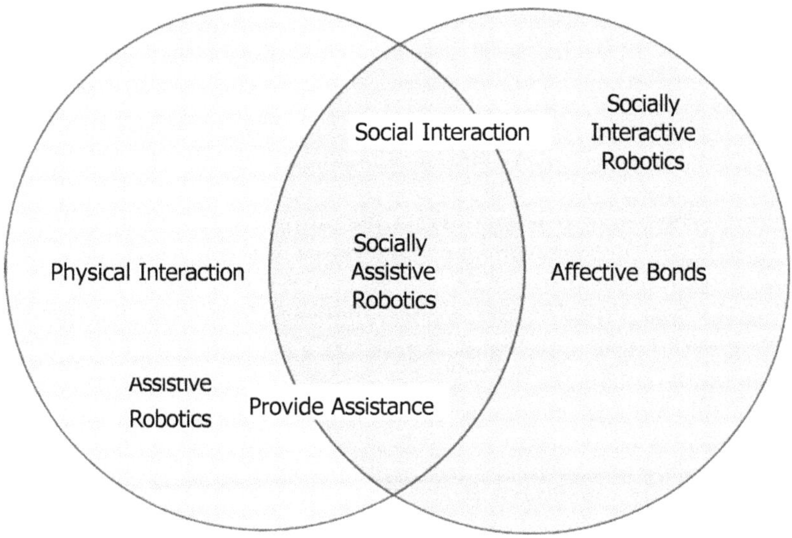

Figure 12.5 Socially assistive robotics: bridging physical assistance with social interaction.

SAR capitalizes on nuanced human-robot interaction (HRI) to innovate therapeutic practices, critically depending on the robots being perceived as trustworthy and interactive. The efficacy of SAR is deeply intertwined with the robot's physical form and, more significantly, its behavior that should replicate social cues and human-like expressions to which humans instinctively respond [63]. SAR platforms are crafted to engage patients in rehabilitation with expressive gestures, gaze, and body language-essential components of nonverbal communication [64, 65]. These platforms employ sophisticated robot heads and arms that provide expressivity without physical manipulation. Additionally, the inclusion of versatile torsos and mobile bases in robots enhances their potential for interactive expressivity, and affordability is a key consideration for their widespread application.

SAR's impact is evident across rehabilitative contexts, from aiding children in motor skill development to coaching stroke patients, improving their motor skills through enjoyable activities such as gaming, which can boost social interaction and elicit positive emotional responses [67, 68]. The field has witnessed growth from companion robots for the elderly to robotic coaches for stroke patients, expanding its scope to aid children with conditions such as CP or OBPP [69]. Examples of SAR's reach include Kinetron's video game-based activities and QTrobot's physical therapy guidance, which have garnered positive feedback for encouraging active rehabilitation participation [70, 71]. SAR systems, including the anthropomorphic Marko and the popular NAO robot, are continually refined through collaborative design, enhancing their interactive capabilities [72, 73]. Figure 12.6 illustrate this. The advancements in SAR underscore the significance of tailoring interactions to enrich the therapeutic experience, aiming to boost engagement and therapeutic outcomes with adaptable feedback and personalized interactions [74, 75].

The autonomous operation of SAR is exemplified by the NT cognitive software architecture, as shown in Figure 12.7, which autonomously conducts interactive rehabilitation sessions, as discussed in [7]. This system bolsters the rehabilitation process with games that promote accurate imitation, offering corrective feedback and rewards for successful exercises. Its architecture facilitates direct patient-robot engagement, with therapists customizing sessions via an intuitive interface and engineers prepping the system remotely, allowing for independent robot operation during therapy. This setup paves the way for SAR to become a self-sufficient yet crucial aspect of rehabilitation, integrating advanced HRI into therapeutic regimens.

Furthermore, the Lil'Flo SAR system, as introduced in [76], is specifically tailored for young patients with cerebral palsy and brachial plexus injuries. Integrating telepresence, cutting-edge computer vision, and humanoid robotic elements, Lil'Flo provides a dynamic and immersive rehabilitation environment. As illustrated in Figure 12.8, the system serves as an intermediary, effectively closing the distance between patients and clinicians during

Figure 12.6 NAO: An autonomous, programmable humanoid robot [66].

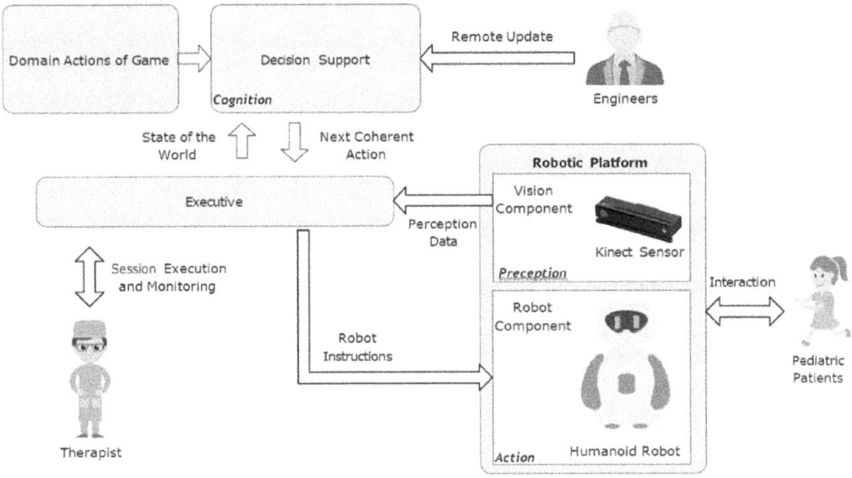

Figure 12.7 An architectural overview of SAR cognitive software.

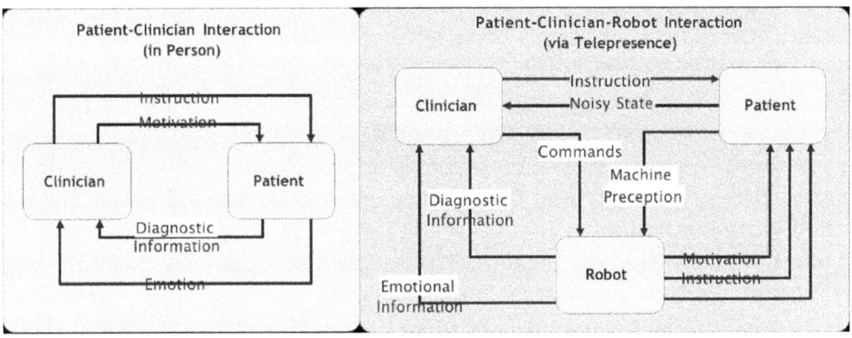

Figure 12.8 A comparison of interactions between patients and clinicians in-person and patients, clinicians, and the robot in telepresence interactions. The diagram illustrates how the inclusion of the robot in telepresence sessions preserves vital communication channels that are otherwise diminished compared to traditional in-person engagements.

remote sessions and preserving the interactive nuances of direct human contact. The system supports active game-play and guided exercises, all under the remote guidance of healthcare professionals. Virtual usability tests by rehabilitation experts have demonstrated Lil'Flo's capacity to enrich therapy experiences, validating the role of SAR in improving patient engagement and participation in therapeutic activities.

12.11 CHALLENGES OF REHABILITATION ROBOTICS

Rehabilitation robotics represents a significant leap forward in patient care, offering innovative approaches to physical therapy. However, the field is not without substantial challenges. The first and perhaps most prohibitive is cost. The development, manufacturing, and maintenance of these advanced systems entail considerable expense, limiting their availability primarily to well-funded healthcare facilities. The sophisticated nature of these robots also necessitates specialized training for operation and maintenance, imposing additional demands on healthcare staff and resources.

Technical and safety concerns constitute another major challenge. Ensuring the safe interaction between machines and humans requires exhaustive testing and adherence to strict regulatory standards, which can be both time-intensive and costly. This safety assurance is critical, as the consequences of malfunctions could be severe for vulnerable patients. Furthermore, the customization required to meet individual patient needs calls for a highly adaptable design, which can be difficult to achieve given the wide range of conditions and impairments that rehabilitation robotics seeks to address.

Integrating these robots into existing therapeutic regimes introduces further complexities. Therapists must be trained not only in the mechanical aspects of the technology but also in how to integrate it into a patient's therapy plan effectively. Moreover, patient engagement with the technology is essential; the robotic systems must be approachable and user-friendly to ensure patient comfort and compliance, which are crucial for successful rehabilitation outcomes.

The demonstration of efficacy is vital in justifying the investment in rehabilitation robotics. Robust clinical evidence is required to establish these systems as beneficial compared to or as a complement to traditional therapeutic methods. This necessitates rigorous research and clinical trials to build a compelling case for their adoption.

Lastly, ethical considerations and data security present ongoing challenges. The rise of robotics in healthcare brings forth questions regarding patient autonomy, the potential decrease in human interaction, and the implications of data privacy. Ensuring the ethical use of robotics and the protection of sensitive patient data remains a priority as the field evolves.

In summary, rehabilitation robotics faces economic barriers, technical and safety issues, challenges in clinical integration, the need for evidence of clinical effectiveness, and ethical concerns. These challenges must be addressed through multidisciplinary collaboration and innovation to fully realize the potential of robotics in enhancing patient rehabilitation.

REFERENCES

1. A. Cieza, K. Causey, K. Kamenov, S. W. Hanson, S. Chatterji, and T. Vos, "Global estimates of the need for rehabilitation based on the global burden of disease study 2019: A systematic analysis for the Global Burden of Disease Study 2019," *Lancet*, vol. 396, no. 10267, pp. 2006–2017, 2020.
2. A. W. Heinemann, M. Feuerstein, W. R. Frontera, S. A. Gard, L. A. Kaminsky, S. Negrini, L. G. Richards, and C. Vallée, "Rehabilitation is a global health priority," *Am J Occup Ther.*, vol. 74, no. 2, pp. 7402170010p1–7402170010p3, 2020. doi: 10.5014/ajot.2020.742006. PMID: 32204770.
3. Y.-S. Jee, "Application of metaverse technology to exercise rehabilitation: Present and future," *J. Exerc. Rehabil.*, vol. 19, no. 2, p. 93, 2023.
4. G. Lenz, "Rehabilitation delivery," *Encycl. Public Health*, Springer, pp. 1243–1246, 2008.
5. E. M. Lund, A. J. Forber-Pratt, C. Wilson, and L. R. Mona, "The COVID-19 pandemic, stress, and trauma in the disability community: A call to action," *Rehabil. Psychol.*, vol. 65, no. 4, p. 313, 2020.
6. Z.-Q. Yang, D. Du, X.-Y. Wei, and R. K.-Y. Tong, "Augmented reality for stroke rehabilitation during COVID-19," *J. Neuroeng. Rehabil.*, vol. 19, no. 1, p. 136, 2022.
7. J. C. Pulido, C. Suarez-Mejias, J. C. Gonzalez, A. D. Ruiz, P. F. Ferri, M. E. M. Sahuquillo, C. E. R. De Vargas, P. Infante-Cossio, C. L. P. Calderon, and F. Fernandez, "A socially assistive robotic platform for upper-limb rehabilitation: A longitudinal study with pediatric patients," *IEEE Robot. Autom. Mag.*, vol. 26, no. 2, pp. 24–39, 2019.
8. B. Butz, A. Jussen, A. Rafi, G. Lux, and J. Gerken, "A taxonomy for augmented and mixed reality applications to support physical exercises in medical rehabilitation—a literature review," *Healthcare*, vol. 10, no. 4, p. 646, 2022.
9. S. C. Leong, Y. M. Tang, F. M. Toh, and K. N. Fong, "Examining the effectiveness of virtual, augmented, and mixed reality (VAMR) therapy for upper limb recovery and activities of daily living in stroke patients: A systematic review and meta-analysis," *J. Neuroeng. Rehabil.*, vol. 19, no. 1, pp. 1–20, 2022.
10. M. Milewska-Je˛drzejczak and A. Gła˛bin´ski, "The influence of conventional and innovative rehabilitation methods on brain plasticity induction in patients with multiple sclerosis," *J. Clin. Med.*, vol. 12, no. 5, p. 1880, 2023.
11. H. Mousavi Hondori and M. Khademi, "A review on technical and clinical impact of Microsoft Kinect on physical therapy and rehabilitation," *J. Med. Eng.*, vol. 2014, 2014.
12. A. Elor, M. Teodorescu, and S. Kurniawan, "Project star catcher: A novel immersive virtual reality experience for upper limb rehabilitation," *ACM Trans. Access. Comput. (TACCESS)*, vol. 11, no. 4, pp. 1–25, 2018.
13. M. F. Pereira, C. Prahm, J. Kolbenschlag, E. Oliveira, and N. F. Rodrigues, "Application of AR and VR in hand rehabilitation: A systematic review," *J. Biomed. Inform.*, vol. 111, p. 103584, 2020.

14. D. Perez-Marcos, O. Chevalley, T. Schmidlin, G. Garipelli, A. Serino, P. Vuadens, T. Tadi, O. Blanke, and J. d. R. Millán, "Increasing upper limb training intensity in chronic stroke using embodied virtual reality: A pilot study," *J. Neuroeng. Rehabil.*, vol. 14, pp. 1–14, 2017.
15. K. N. Fong, Y. M. Tang, K. Sie, A. K. Yu, C. C. Lo, and Y. W. Ma, "Task-specific virtual reality training on hemiparetic upper extremity in patients with stroke," *Virtual Reality*, vol. 26, pp. 453–464, 2022.
16. M. S. Cameirão, S. Bermúdez i Badia, and P. F. Verschure, "Virtual reality-based upper extremity rehabilitation following stroke: A review," *J. CyberTherapy Rehabil.*, vol. 1, no. 1, pp. 63–74, 2008.
17. G. Saposnik and M. Levin, "Virtual reality in stroke rehabilitation: A meta-analysis and implications for clinicians," *Stroke*, vol. 42, no. 5, pp. 1380–1386, 2011.
18. J. Iruthayarajah, A. McIntyre, A. Cotoi, S. Macaluso, and R. Teasell, "The use of virtual reality for balance among individuals with chronic stroke: A systematic review and meta-analysis," *Top. Stroke Rehabil.*, vol. 24, no. 1, pp. 68–79, 2017.
19. M. Maier, B. Rubio Ballester, A. Duff, E. Duarte, and P. F. Verschure, "Effect of specific over nonspecific VR-based rehabilitation on poststroke motor recovery: A systematic meta-analysis," *Neurorehabil Neural Repair,* vol. 33, no. 2, pp. 112–129. doi: 10.1177/1545968318820169. Epub 2019 Jan 30. PMID: 30700224; PMCID: PMC6376608.
20. C. Gómez-Portes, D. Carneros-Prado, J. Albusac, J. J. Castro-Schez, C. Glez-Morcillo, and D. Vallejo, "Phyre up! A system based on mixed reality and gamification to provide home rehabilitation for stroke patients," *IEEE Access*, vol. 9, pp. 139122–139137, 2021.
21. M. C. Howard and M. M. Davis, "A meta-analysis and systematic literature review of mixed reality rehabilitation programs: Investigating design characteristics of augmented reality and augmented virtuality," *Comput. Hum. Behav.*, vol. 130, p. 107197, 2022.
22. M. Hudák, B. Sobota, Š. Korečko, and M. Sivy, "Fully immersive web-based collaborative virtual environment for upper limb rehabilitation purposes," in *2020 18th International Conference on Emerging eLearning Technologies and Applications (ICETA)*, IEEE, 2020, pp. 194–199.
23. S. Mann, T. Furness, Y. Yuan, J. Iorio, and Z. Wang, "All reality: Virtual, augmented, mixed (X), mediated (X, Y), and multimediated reality," *arXiv* preprint arXiv:1804.08386, 2018.
24. D. L. Jaffe, D. A. Brown, C. D. Pierson-Carey, E. L. Buckley, and H. L. Lew, "Stepping over obstacles to improve walking in individuals with poststroke hemiplegia," *J. Rehabil. Res. Dev.*, vol. 41, pp. 283–292, 2004. doi: 10.1682/jrrd.2004.03.0283. PMID: 15543446.
25. Y. M. Aung and A. Al-Jumaily, "Augmented reality-based illusion system with biofeedback," in *2nd Middle East Conference on Biomedical Engineering*, IEEE, 2014, pp. 265–268.
26. Y. M. Aung and A. Al-Jumaily, "Augmented reality-based Rehabio system for shoulder rehabilitation," *Int. J. Mechatronics Autom.*, vol. 4, no. 1, pp. 52–62, 2014.

27. N. De Leon, S. Bhatt, and A. Al-Jumaily, "Augmented reality game-based multi-usage rehabilitation therapist for stroke patients," *Int. J. Smart Sensing Intell. Syst.*, vol. 7, no. 3, pp. 1044–1058, 2014.
28. C.-H. Lee, Y. Kim, and B.-H. Lee, "Augmented reality-based postural control training improves gait function in patients with stroke: Randomized controlled trial," *Hong Kong Physiother. J.*, vol. 32, no. 2, pp. 51–57, 2014.
29. J. Trojan, M. Diers, X. Fuchs, F. Bach, R. Bekrater-Bodmann, J. Foell, S. Kamping, M. Rance, H. Maaß, and H. Flor, "An augmented reality home-training system based on the mirror training and imagery approach," *Behav. Res. Methods*, vol. 46, pp. 634–640, 2014.
30. A. G. D. Corrêa, G. A. de Assis, M. do Nascimento, and R. de Deus Lopes, "Perceptions of clinical utility of an augmented reality musical software among health care professionals," *Disabil. Rehabil. Assist. Technol.*, vol. 12, no. 3, pp. 205–216, 2017.
31. S. Ponte, S. Gabrielli, J. Jonsdottir, M. Morando, and S. Dellepiane, "Monitoring game-based motor rehabilitation of patients at home for better plans of care and quality of life," in *2015 37th Annual International Conference of the IEEE Engineering in Medicine and Biology Society (EMBC)*, IEEE, 2015, pp. 3941–3944.
32. J. Collins, S. Hoermann, and H. Regenbrecht, "Comparing a finger dexterity assessment in virtual, video-mediated, and unmediated reality," *Int. J. Child Health Hum. Dev.*, vol. 9, no. 3, pp. 333–342, 2016.
33. A. E. F. Da Gama, T. M. Chaves, L. S. Figueiredo, A. Baltar, M. Meng, N. Navab, V. Teichrieb, and P. Fallavollita, "Mirrarbilitation: A clinically-related gesture recognition interactive tool for an AR rehabilitation system," *Comput. Methods Programs Biomed.*, vol. 135, pp. 105–114, 2016.
34. C. Colomer, R. Llorens, E. Noé, and M. Alcañiz, "Effect of a mixed reality-based intervention on arm, hand, and finger function on chronic stroke," *J. Neuroengineering Rehabil.*, vol. 13, no. 1, pp. 1–11, 2016.
35. J. Liu, J. Mei, X. Zhang, X. Lu, and J. Huang, "Augmented reality-based training system for hand rehabilitation," *Multimed. Tools Appl.*, vol. 76, pp. 14,847–14,867, 2017.
36. S. Hoermann, L. Ferreira dos Santos, N. Morkisch, K. Jettkowski, M. Sillis, H. Devan, P. S. Kanagasabai, H. Schmidt, J. Krüger, C. Dohle et al., "Computerized mirror therapy with augmented reflection technology for early stroke rehabilitation: clinical feasibility and integration as an adjunct therapy," *Disabil. Rehabil.*, vol. 39, no. 15, pp. 1503–1514, 2017.
37. J. Pinches and S. Hoermann, "Automated instructions and real-time feedback for upper limb computerized mirror therapy with augmented reflection technology," *J. Altern. Med. Res.*, vol. 10, no. 1, pp. 37–46, 2018.
38. W.-W. Liao, S. McCombe Waller, and J. Whitall, "Kinect-based individualized upper extremity rehabilitation is effective and feasible for individuals with stroke using a transition from clinic to home protocol," *Cogent Med.*, vol. 5, no. 1, p. 1428038, 2018.
39. C. M. An and D. H. Kim, "Clinical application of AR system in early rehabilitation program after stroke: 2 case study," *J. Korean Phys. Ther.*, vol. 31, no. 3, pp. 141–146, 2019.

40. A. Dash, A. Yadav, A. Chauhan, and U. Lahiri, "Kinect-assisted performance-sensitive upper limb exercise platform for post-stroke survivors," *Front. Neurosci.*, vol. 13, p. 228, 2019.
41. A. Adomavičienė, K. Daunoravičienė, R. Kubilius, L. Varžaitytė, and J. Raistenskis, "Influence of new technologies on post-stroke rehabilitation: a comparison of Armeo Spring to the Kinect system," *Medicina*, vol. 55, no. 4, p. 98, 2019.
42. A. Orand, E. Erdal Aksoy, H. Miyasaka, C. Weeks Levy, X. Zhang, and C. Menon, "Bilateral tactile feedback-enabled training for stroke survivors using Microsoft Kinect™," *Sensors*, vol. 19, no. 16, p. 3474, 2019.
43. E. R. Høeg, J. R. Bruun-Pedersen, S. Cheary, L. K. Andersen, R. Paisa, S. Serafin, and B. Lange, "Buddy biking: a user study on social collaboration in a virtual reality exergame for rehabilitation," *Virtual Real.*, vol. 27, no. 1, pp. 245–262, 2023.
44. A. M. Edwards, L. Dutton-Challis, D. Cottrell, J. H. Guy, and F. J. Hettinga, "Impact of active and passive social facilitation on self-paced endurance and sprint exercise: encouragement augments performance and motivation to exercise," *BMJ Open Sport Exerc. Med.*, vol. 4, no. 1, p. e000368, 2018.
45. C. Anderson-Hanley, A. L. Snyder, J. P. Nimon, and P. J. Arciero, "Social facilitation in virtual reality-enhanced exercise: competitiveness moderates exercise effort of older adults," *Clin. Interv. Aging*, pp. 275–280, 2011.
46. G. Alankus, R. Proffitt, C. Kelleher, and J. Engsberg, "Stroke therapy through motion-based games: a case study," *ACM Trans. Accessible Comput. (TACCESS)*, vol. 4, no. 1, pp. 1–35, 2011.
47. M. Hudák and M. Sivy, "Web-based collaborative virtual environments to support cross-platform access," in *Poster 2019 International Student Scientific Conference*, IEEE, 2019, pp. 178–182. doi: 10.1109/ICETA51985.2020.9379239.
48. F. Cerritelli, M. Chiera, M. Abbro, V. Megale, J. Esteves, A. Gallace, and A. Manzotti, "The challenges and perspectives of the integration between virtual and augmented reality and manual therapies," *Front. Neurol.*, vol. 12, p. 700211, 2021.
49. V. S. Huang and J. W. Krakauer, "Robotic neurorehabilitation: a computational motor learning perspective," *J. Neuroengineering Rehabil.*, vol. 6, pp. 1–13, 2009.
50. K. Moulaei, K. Bahaadinbeigy, A. A. Haghdoost, M. S. Nezhad, and A. Sheikhtaheri, "Overview of the role of robots in upper limb disabilities rehabilitation: a scoping review," *Arch. Public Health*, vol. 81, no. 1, p. 84, 2023.
51. M. Gilliaux, A. Renders, D. Dispa, D. Holvoet, J. Sapin, B. Dehez, C. Detrembleur, T. M. Lejeune, and G. Stoquart, "Upper limb robot-assisted therapy in cerebral palsy: a single-blind randomized controlled trial," *Neurorehabil. Neural Repair*, vol. 29, no. 2, pp. 183–192, 2015.
52. Q. Wu and H. Wu, "Development, dynamic modeling, and multi-modal control of a therapeutic exoskeleton for upper limb rehabilitation training," *Sensors*, vol. 18, no. 11, p. 3611, 2018.
53. N. Schweighofer, Y. Choi, C. Winstein, and J. Gordon, "Task-oriented rehabilitation robotics," *Am. J. Phys. Med. Rehabil.*, vol. 91, no. 11, pp. S270–S279, 2012.

54. M. J. Johnson, R. C. Loureiro, and W. S. Harwin, "Collaborative tele-rehabilitation and robot-mediated therapy for stroke rehabilitation at home or clinic," *Intell. Serv. Robotics*, vol. 1, no. 2, pp. 109–121, 2008.
55. S. Bhujel and S. Hasan, "A comparative study of end-effector and exoskeleton type rehabilitation robots in human upper extremity rehabilitation," *Human-Intell. Syst. Integr.*, vol. 5, no. 1, pp. 11–42, 2023.
56. L. Zhang, S. Guo, and Q. Sun, "Development and assist-as-needed control of an end-effector upper limb rehabilitation robot," *Appl. Sci.*, vol. 10, no. 19, p. 6684, 2020.
57. D. Su, Z. Hu, J. Wu, P. Shang, and Z. Luo, "Review of adaptive control for stroke lower limb exoskeleton rehabilitation robot based on motion intention recognition," *Front. Neurorobotics*, vol. 17, 2023. doi: 10.3389/fnbot.2023.1186175. PMID: 37465413; PMCID: PMC10350518.
58. S. Jacob, M. Alagirisamy, C. Xi, V. Balasubramanian, R. Srinivasan, R. Parvathi, N. Jhanjhi, and S. M. Islam, "AI and IoT-enabled smart exoskeleton system for rehabilitation of paralyzed people in connected communities," *IEEE Access*, vol. 9, pp. 80,340–80,350, 2021.
59. D. Leonardis, M. Solazzi, I. Bortone, and A. Frisoli, "A 3-RSR haptic wearable device for rendering fingertip contact forces," *IEEE Trans. Haptics*, vol. 10, no. 3, pp. 305–316, 2016.
60. F. Chinello, M. Malvezzi, C. Pacchierotti, and D. Prattichizzo, "Design and development of a 3RRS wearable fingertip cutaneous device," in *2015 IEEE International Conference on Advanced Intelligent Mechatronics (AIM)*, IEEE, 2015, pp. 293–298.
61. I. Bortone, D. Leonardis, N. Mastronicola, A. Crecchi, L. Bonfiglio, C. Procopio, M. Solazzi, and A. Frisoli, "Wearable haptics and immersive virtual reality rehabilitation training in children with neuromotor impairments," *IEEE Trans. Neural Syst. Rehabil. Eng.*, vol. 26, no. 7, pp. 1469–1478, 2018.
62. Á. A. Cuadrado, "Rehabilitaci6n del ACV: evaluaci6n, pron6stico y tratamiento," *Galicia Clínica*, vol. 70, no. 3, pp. 25–40, 2009.
63. J. Goetz, S. Kiesler, and A. Powers, "Matching robot appearance and behavior to tasks to improve human-robot cooperation," in *The 12th IEEE International Workshop on Robot and Human Interactive Communication, 2003. Proceedings. ROMAN 2003*, IEEE, 2003, pp. 55–60.
64. A. Tapus, M. J. Mataric, and B. Scassellati, "Socially assistive robotics [grand challenges of robotics]," *IEEE Robotics Autom. Mag.*, vol. 14, no. 1, pp. 35–42, 2007.
65. A. Specian, R. Mead, S. Kim, M. Mataric´, and M. Yim, "Quori: A community-informed design of a socially interactive humanoid robot," *IEEE Trans. Robotics*, vol. 38, no. 3, pp. 1755–1772, 2021.
66. SoftBank Robotics, "Nao - the humanoid and programmable robot | SoftBank Robotics," [Online]. Available: https://www.aldebaran.com/en/nao. [Accessed: Feb. 10, 2024].
67. B. Robins, E. Ferrari, K. Dautenhahn, G. Kronreif, B. Prazak-Aram, G.-J. Gelderblom, B. Tanja, F. Caprino, E. Laudanna, and P. Marti, "Human-centred design methods: Developing scenarios for robot assisted play informed by user panels and field trials," *Int. J. Hum.-Comput. Stud.*, vol. 68, no. 12, pp. 873–898, 2010.

68. J. Wainer, D. J. Feil-Seifer, D. A. Shell, and M. J. Mataric, "Embodiment and human-robot interaction: A task-based perspective," in *RO-MAN 2007-The 16th IEEE International Symposium on Robot and Human Interactive Communication*, IEEE, 2007, pp. 872–877.
69. N. A. Malik, F. A. Hanapiah, R. A. A. Rahman, and H. Yussof, "Emergence of socially assistive robotics in rehabilitation for children with cerebral palsy: A review," *Int. J. Adv. Robotic Syst.*, vol. 13, no. 3, p. 135, 2016.
70. O. Kachmar, V. Kozyavkin, and I. Ablikova, "Humanoid social robots in the rehabilitation of children with cerebral palsy," in *REHAB 2014*, ICST, 2014.
71. F. J. Rodriguez-Lera, L. Gomes, P. Ziafati, A. Nazarikhorram, A. Stefanetti, and A.-M. Schuller, "Emotional robots for coaching: Motivating physical rehabilitation using emotional robots," in *Proceedings of the Personal Robots for Exercising and Coaching (PREC)*, ACM /IEEE, pp. 1–7, 2018.
72. B. Borovac, M. Gnjatovic´, S. Savic´, M. Rakovic´, and M. Nikolic´, "Human-like robot Marko in the rehabilitation of children with cerebral palsy," in *New Trends in Medical and Service Robots: Assistive, Surgical and Educational Robotics*, Springer, 2016, pp. 191–203.
73. F. Martí Carrillo, J. Butchart, S. Knight, A. Scheinberg, L. Wise, L. Sterling, and C. McCarthy, "Adapting a general-purpose social robot for pediatric rehabilitation through in situ design," *ACM Trans. Hum.-Robot Interact.*, vol. 7, no. 1, pp. 1–30, 2018.
74. M. Fridin and M. Belokopytov, "Robotics agent coacher for CP motor function (RAC CP Fun)," *Robotica*, vol. 32, no. 8, pp. 1265–1279, 2014.
75. J. Greczek, E. Kaszubski, A. Atrash, and M. Mataric´, "Graded cueing feedback in robot-mediated imitation practice for children with autism spectrum disorders," in *The 23rd IEEE International Symposium on Robot and Human Interactive Communication*, IEEE, 2014, pp. 561–566.
76. M. J. Sobrepera, V. G. Lee, and M. J. Johnson, "The design of Lil'Flo, a socially assistive robot for upper extremity motor assessment and rehabilitation in the community via telepresence," *J. Rehabil. Assist. Technol. Eng.*, vol. 8, p. 20556683211001805, 2021.

Chapter 13

Empowering Patient Care through Metaverse Technologies

Md Nurunnabi

13.1 INTRODUCTION

The concept of the metaverse, characterized by persistent, shared, 3D virtual spaces linked into a perceived virtual universe, offers a transformative platform for rehabilitation practices. The integration of metaverse technologies with rehabilitation has ushered in a new era in therapeutic practices, characterized by immersive, interactive environments that simulate real-world activities or create entirely new therapeutic scenarios. This chapter synthesizes findings from past studies, exploring the implementation and efficacy of various technologies such as virtual reality (VR), augmented reality (AR), mixed reality (MR), socially assistive robots (SAR), and cloud-based platforms in the realm of rehabilitation.

The potential of VR in rehabilitation is profound, providing environments that can be meticulously controlled and manipulated to cater to the specific therapeutic needs of individuals. As noted in several studies, VR allows for the creation of detailed simulations tailored to enhance motor function and cognitive recovery, significantly improving patient outcomes and adherence to therapeutic regimes [1–3]. AR and MR further expand these capabilities by integrating digital elements with the real world, offering more dynamic and responsive therapeutic experiences. These technologies not only enhance patient engagement through interactive and augmented overlays but also allow for real-time feedback and adjustment of therapeutic activities [4–6].

Moreover, the rise of SAR introduces a novel dimension to rehabilitation, wherein robots are not merely tools for physical therapy but also act as social interactants that provide emotional support and motivation [7, 8]. These robots are particularly effective in pediatric rehabilitation, where they leverage naturalistic interactions to enhance patient engagement and therapeutic adherence [9–11].

Cloud-based platforms and the integration of the Internet of Things (IoT) stand at the forefront of making rehabilitation services more accessible and efficient. These technologies facilitate remote therapy sessions and real-time monitoring, thus breaking geographical barriers to care and allowing

DOI: 10.1201/9781032663609-16

continuous adaptation of therapy programs based on patient progress [12, 13]. This capability is crucial for providing ongoing and personalized care, particularly in underserved regions or situations where patients cannot frequently visit healthcare facilities.

Despite these technological advances, the integration of metaverse concepts into clinical practice presents challenges, including the high costs associated with advanced technologies, their technical complexity, and the necessity for substantial clinical validation to ensure their effectiveness and safety across diverse patient populations [14–16]. These challenges underscore the need for continued research and development within this field.

As we delve deeper into the applications of these technologies in rehabilitation, this chapter will explore how they not only improve current therapeutic practices but also open up new avenues for patient care, fundamentally transforming the landscape of rehabilitation medicine. This exploration is supported by a robust review of literature and clinical studies that highlight both the potential and the limitations of metaverse technologies in therapeutic settings.

13.2 EVOLUTION OF METAVERSE TECHNOLOGIES IN REHABILITATION

This section provides an in-depth examination of the historical evolution of metaverse technologies within the field of rehabilitation. From the earliest conceptual explorations of VR to the sophisticated integration of augmented and mixed realities, SAR, and cloud-based platforms, each technological breakthrough has significantly influenced rehabilitation practices. The progression of these technologies reflects a broader trend toward creating more immersive, interactive, and patient-centered therapeutic environments that are presented in Table 13.1. This evolution has not only enhanced therapeutic outcomes but also transformed patient experience in rehabilitation settings.

13.3 TECHNOLOGICAL INTEGRATION IN REHABILITATION

The incorporation of metaverse technologies such as VR, AR, MR, SAR, and cloud-based platforms in rehabilitation practices signifies a transformative leap forward in enhancing therapeutic outcomes and patient engagement. This section delves into the specific technologies and equipment that facilitate these advances, providing a comprehensive view of how these components integrate into a cohesive system for metaverse-based rehabilitation.

Table 13.1 Historical development of rehabilitation in the metaverse

Year	*Milestone or Innovation*	*Description*	*References*
1960s	Inception of VR	Early VR systems developed, laying the groundwork for immersive environments.	[17]
1980s	First Use of VR in Therapy	VR used experimentally to treat phobias and anxiety, initiating its therapeutic applications.	[18]
1990s	Advancements in VR Hardware	Development of more accessible and functional VR systems such as the CAVE system and commercial headsets.	[19]
2000s	Introduction of AR	AR starts being integrated into rehabilitation for enhancing physical therapy through interactive overlays.	[20]
2010	Socially Assistive Robots (SAR)	Deployment of robots such as NAO and Pepper in therapeutic settings to provide social interaction and support.	[21]
2012	Cloud-Based Rehabilitation Platforms	First implementations of cloud technology to enhance the accessibility and effectiveness of rehab therapies.	[22]
2014	Integration of IoT in Rehabilitation	Incorporation of IoT devices for continuous monitoring and adaptive rehabilitation programs.	[23]
2016	MR Systems in Rehab	Introduction of systems such as Microsoft HoloLens to blend real and virtual for more engaging rehab activities.	[24]
2018	AI-Driven Rehabilitation Platforms	AI technologies begin to be integrated into rehabilitation to provide personalized therapy adjustments based on real-time data.	[25]
2020	Metaverse Concept in Mainstream Rehabilitation	The term "metaverse" is increasingly used to describe integrated VR, AR, and MR systems in rehabilitation contexts.	[26]
2022	Advanced VR and AR Therapeutic Protocols	Development of specific protocols and applications in VR and AR tailored for rehabilitation in diverse patient populations.	[27]

13.3.1 Virtual and Augmented Realities in Rehabilitation

Virtual and augmented realities are at the forefront of this technological revolution, creating immersive and interactive environments that significantly enhance motor and cognitive rehabilitation outcomes. Essential equipment and technologies typically used in VR and AR setups include the following:

- **Head-Mounted Displays (HMDs):** Devices such as the Oculus Rift, HTC Vive, or Microsoft HoloLens that provide immersive visual and auditory experiences.
- **Motion Tracking Sensors:** Systems such as the Leap Motion, Kinect sensors, or VR treadmills that track body movements and translate them into the virtual environment to support tasks ranging from basic movements to complex rehabilitation exercises.
- **Haptic Feedback Devices:** These include gloves or suits such as the Teslasuit or HaptX Gloves, which provide tactile feedback to users, enhancing the realism of virtual tasks and exercises by simulating physical sensations.

Table 13.2 shows the hardware and software components of the immersive VR rehabilitation. Studies such as those by Gotsis et al. [1] and Aviv Elor et al. [3] demonstrate the efficacy of these technologies in providing environments that not only aid in the recovery of physical abilities but also engage patients in cognitive tasks, significantly boosting therapy adherence and effectiveness.

13.3.2 Role of SAR

SAR play a crucial role in personalized patient care within the metaverse framework. These robots, which include platforms such as NAO and Pepper, are equipped with features such as:

- **Emotive AI Systems:** These systems analyze and respond to human emotions, enhancing interaction.
- **Voice and Facial Recognition Technologies:** These systems enable personalized communication and monitoring, as discussed in studies by Sobrepera et al. [9] and Pulido et al. [11].
- **Mobile Bases and Manipulators:** These systems allow robots to assist with physical tasks and exercises, adapting to patient needs in real time.

These robots provide not just physical rehabilitation support but also crucial emotional and motivational engagement, which is essential for long-term therapy success.

Table 13.2 Hardware and software components of immersive VR rehabilitation (IVRR) system

Component Type	*Component Name*	*Description*	*Function in System*
Hardware	VR Headset (e.g., Oculus Rift, HTC Vive)	A device worn on the head that displays immersive virtual environments and tracks head movements.	Provides immersive visual and auditory feedback, allowing the patient to interact with the virtual rehabilitation environment.
	Full-body Motion-Tracking Suit	A wearable suit embedded with sensors that capture the movements of various body parts.	Enables precise monitoring of patient movements during exercises, crucial for accurate therapy assessment and adjustment.
	Motion-Tracking Gloves	Gloves equipped with sensors to capture detailed hand and finger movements.	Enhances the tracking of fine motor skills and hand movements during rehabilitation exercises.
	Treadmill with Integrated Sensors	A treadmill equipped with sensors that measure speed, gait, and other relevant parameters.	Used for gait training and balance exercises, providing data crucial for tailoring exercises to patient needs.
	Digital Display Screen	A screen that displays real-time feedback, instructions, and motivational messages.	Acts as a socially assistive interface, offering real-time encouragement and detailed guidance to enhance the patient's performance.
Software	VR Rehabilitation Software	Custom software that runs on the VR headset, rendering the virtual environment and rehabilitation scenarios.	Manages the virtual scenes and interactive exercises, providing a controlled yet dynamic setting for therapy.
	Motion Analysis and Feedback Software	Software that processes data from motion-tracking hardware to analyze patient performance.	Provides real-time feedback based on motion data, allowing for immediate corrective suggestions and adjustments.
	Cloud-based Data Management System	A platform that stores, processes, and manages therapy data and patient progress in the cloud.	Facilitates remote monitoring by healthcare professionals and integrates data across sessions for comprehensive management.
	Gamification Software	Software that integrates game mechanics into the rehabilitation exercises.	Enhances patient engagement by incorporating elements such as points, levels, and rewards into the therapy sessions.

13.3.3 Cloud-Based Platforms and IoT

Cloud-based platforms and the IoT are pivotal in extending the reach and efficacy of rehabilitation services. These technologies ensure seamless integration of various components such as:

- **Wearable Sensors:** Devices worn by patients that collect data on movement patterns, physiological metrics, and rehabilitation progress.
- **Cloud Computing Platforms:** Such as Amazon Web Services or Microsoft Azure, which process and store data, allowing for real-time feedback and adjustment of rehabilitation protocols.
- **High-Speed Internet Connections:** Ensure uninterrupted data transmission between local devices and cloud servers, facilitating a smooth and responsive rehabilitation experience.

As highlighted in studies by Hudák et al. [12] and Inamura and Mizuchi [13], these technologies are crucial for enabling remote rehabilitation services, enhancing the accessibility of personalized treatment plans, and providing continuous patient care, critical for widespread adoption of metaverse-based rehabilitation.

13.4 DETAILED METHODS OF REHABILITATION USING METAVERSE TECHNOLOGIES

The integration of metaverse technologies in rehabilitation systems involves an intricate combination of hardware and software to create immersive, interactive, and therapeutic environments tailored to individual patient needs. This section explores various methods of rehabilitation using metaverse technologies, elaborating on the essential hardware and software, and providing a detailed view of their integration.

13.4.1 Metaverse-Based Therapeutic Rehabilitation (MBTR)

a. **Hardware and Software Components:** Below is a detailed table (Table 13.1) outlining the hardware and software components utilized in the MBTR system, designed to enhance physical rehabilitation through VR. This table describes each component's purpose and function within the system:
b. **Methodology of MBRT:** Figure 13.1 depicts an advanced metaverse-based rehabilitation system where multiple patient A and B are immersed in a virtual environment for therapeutic exercises. The patients A and B, outfitted with a VR headset and full-body motion-tracking gear, including gloves and a suit, is engaged in a simulated

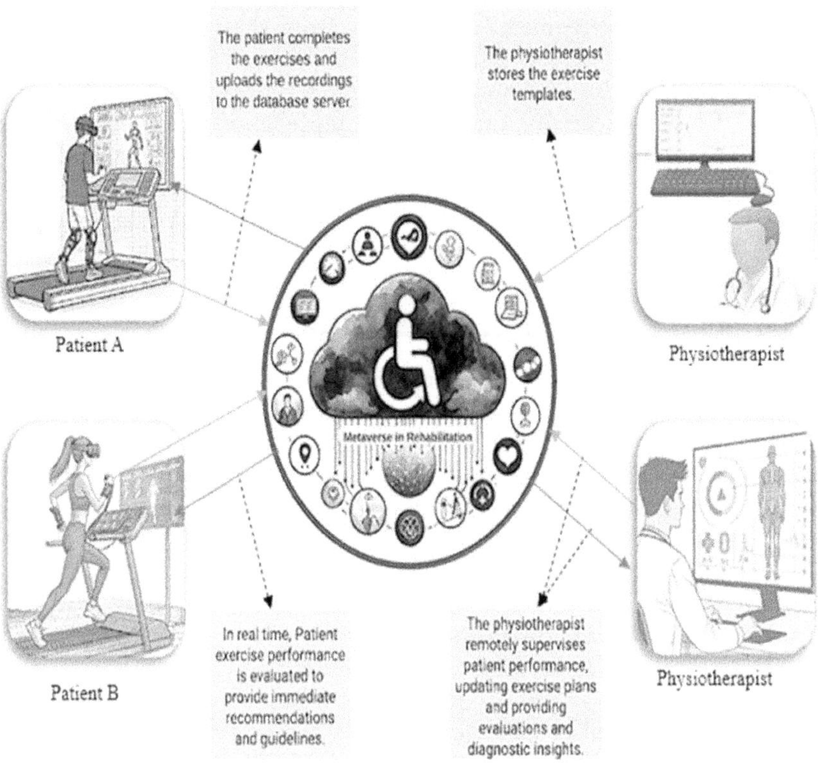

Figure 13.1 Metaverse-based therapeutic rehabilitation (MBTR).

setting designed to mimic real-world activities, such as reaching for or manipulating virtual objects. This setup is part of a rehabilitation exercise aimed at improving motor skills and coordination. Beside the patient, a digital screen functions as a socially assistive system, displaying real-time encouragement and detailed exercise guidance to enhance the patient's performance. This interactive approach not only motivates but also ensures the exercises are performed accurately, maximizing therapeutic benefits.

The environment includes a treadmill integrated with sensors and safety harnesses, used for specific rehabilitation tasks such as gait training or balance exercises. These sensors collect data on the patient's movements, which is crucial for tailoring the virtual experience and exercise difficulty based on real-time performance. Above this setup, a cloud symbol visually represents the cloud computing infrastructure of the system, illustrating the seamless integration of data analytics, patient data storage, and feedback loops. This cloud interface ensures that all data collected from the patient's equipment and sensors is processed in real time, enabling immediate adjustments to the therapy regimen. Data flows from the patient's sensors and the interactive screen into the cloud, depicted through green lines, emphasizing the continuous, real-time data transmission and analysis. This connectivity supports a comprehensive monitoring system where a remote healthcare professional can observe and manage the rehabilitation process. The professional, shown on the right side of Figure 13.1, analyzes the data and adjusts as needed, sending feedback directly to the patient's VR system.

13.4.2 AR-Enhanced Physical Therapy

To develop a home-based rehabilitation system for stroke patients that leverages mixed reality (MR) and gamification techniques to enhance patient engagement and therapeutic effectiveness, while enabling therapists to remotely monitor progress and customize treatment plans. Table 13.3 shows the AR-Enhance Physcial therapy hardware and software component.

Hardware and Software Components

a. **Integration of Hardware and Software**
b. **Data Flow and Processing**

The Azure Kinect sensor captures real-time movement data of the patient performing prescribed exercises in Figure 13.2. This data is processed to track the accuracy and quality of exercise execution. Movement data is then sent to the HoloLens for real-time feedback.

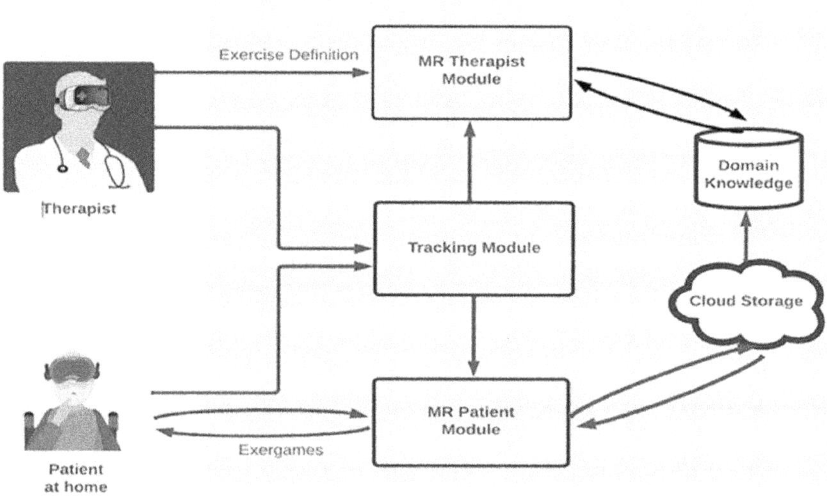

Figure 13.2 General overview and key components of the ARPT architecture [05].

Table 13.3 Hardware and software components of the ARPT system

Component Type	*Component Name*	*Description*	*Function in System*
Hardware	Microsoft HoloLens 2	An advanced MR headset that overlays digital information onto the real world.	Used by therapists to create exercises and by patients to see and interact with rehabilitation exercises in their physical environment.
	Azure Kinect DK	A sensor device that provides high-fidelity spatial awareness and motion tracking.	Tracks the patient's physical movements to ensure exercises are performed correctly and to collect data for analysis.
Software	MR Software for HoloLens 2	Custom-developed application running on the HoloLens 2, displaying interactive 3D content and exercises.	Displays the rehabilitation exercises in MR, providing interactive, visual guides and feedback to the patient.
	Tracking and Analysis Software	Operates with the Azure Kinect to capture and analyze motion data.	Processes the tracking data to monitor exercise accuracy and patient performance, adjusting feedback accordingly.
	Cloud-based Backend (Microsoft Azure)	Hosts the application's backend, storing data and managing system configurations.	Stores exercise data, patient progress, and allows therapists to remotely access and update rehabilitation plans.
	Rehabilitation Planning Software	Allows therapists to define and customize exercises and routines visually.	Facilitates the creation of personalized rehabilitation programs, utilizing MR to visualize and adjust exercises in real-time.
	Gamification Software	Integrates game elements into the rehabilitation exercises.	Engages patients during rehabilitation sessions by incorporating scoring, levels, and motivational elements into the exercises.

- **MR and Exercise Visualization:**

Therapists use HoloLens to create and visualize rehabilitation exercises in a 3D space, setting up virtual markers and paths that patients must follow. Patients wearing the HoloLens see these exercises overlaid in their physical environment, guiding them through each movement with visual cues and corrections.

c. **System Operation and Workflow:**

- **Therapist Module:**

Therapists log into a secure interface via HoloLens or a PC to create and modify patient-specific rehabilitation routines using natural user interfaces (voice, gestures). Exercises are defined using the MR therapist's module, which allows therapists to visually set up the trajectories and key points that patients must reach during their exercises.

- **Patient Module:**

Patients perform exercises at home with the HoloLens on, following the virtual cues displayed in their environment. The system uses gamification elements (e.g., scoring, levels) integrated into the MR environment to make the rehabilitation process more engaging.

- **Feedback and Adjustment:**

As patients perform exercises, the system provides instant visual and auditory feedback via HoloLens based on their performance, which is monitored by Azure Kinect. The system automatically adjusts exercise difficulty and provides corrective feedback to ensure optimal performance and progression. Continuous data collection via Azure Kinect feeds into the tracking module, which analyzes joint movements and overall exercise execution. This data is stored and processed on Microsoft Azure, where therapists can review progress, adjust exercise programs, and provide additional guidance remotely. All patient data and exercise configurations are securely stored in the cloud (Microsoft Azure), ensuring data integrity and confidentiality. The system adheres to healthcare compliance and data protection standards to safeguard sensitive information.

The ARPT system represents a sophisticated integration of MR technology, advanced tracking hardware, and interactive software components, all designed to provide an engaging, effective, and adaptable home rehabilitation environment for stroke patients. By combining MR with gamification and robust data tracking, the system not only enhances the therapeutic experience but also empowers therapists to deliver personalized, data-driven rehabilitation remotely.

d. Mathematical Modeling and Algorithms in ARPT

The ARPT system employs advanced tracking and interactive AR technologies to guide stroke patients through prescribed rehabilitation exercises. This section describes the algorithms and mathematical models that underpin the operation of the system, focusing particularly on the motion tracking and feedback components.

Algorithm 1: Pose Accuracy Algorithm
The Pose Accuracy Algorithm is pivotal in ensuring that the patient performs the exercises with the correct posture and alignment. The Azure Kinect DK sensor captures the 3D positions and orientations of the patient's joints, which are then compared to a predefined "ideal" pose using the following mathematical model:

Euclidean Distance Calculation:
This is a measure of the straight-line distance between two points in Euclidean space. Here, it measures the distance between the actual joint position (P_i) and the ideal joint position (P_i^*).

$$d_i = \sqrt{\left(P_{ix} - P_{ix}^*\right)^2 + \left(P_{iy} - P_{iy}^*\right)^2 + \left(P_{iz} - P_{iz}^*\right)^2} \tag{13.1}$$

where,

P: The position vector of joint i in the patient's current pose by the patient during the rehabilitation session.
P_i*: The position vector of joint i in the ideal pose that the patient is supposed to achieve.
n: Total number of joints being tracked.

These vectors help in determining how close or far the patient's current posture is from the desired posture. This distance is used to calculate how accurately the patient is performing the exercise relative to the ideal pose.

Overall Pose Accuracy Score: This score is calculated to evaluate the patient's pose accuracy.

$$\text{Accuracy Score} = \frac{1}{n}\sum_{i=1}^{n} e^{-\alpha d_i} \tag{13.2}$$

Here, α is a scaling factor that adjusts how much the distance (d_i) affects the score. A higher distance will result in a lower accuracy score. This score is calculated in real time and is used to provide immediate feedback to the patient through the Microsoft HoloLens 2, guiding them to adjust their pose to match the ideal parameters.

Algorithm 2: Rehabilitation Progress Tracker
This algorithm assesses the patient's progress over multiple sessions. It uses a series of performance metrics gathered during each exercise session to update a model that predicts recovery trends.

Progress Calculation:

$$R_t = \frac{1}{n}\sum_{t=1}^{T} S_t \tag{13.3}$$

$$\text{Progress Percentage} = \left(\frac{R_t}{G}\right) \times 100\% \tag{13.4}$$

here:

S_t : The session score for session t.
T: Total number of completed sessions.
G: Target goal for rehabilitation progress.

Linear Regression Model for Trend Analysis:
Regression analysis is used to project future recovery paths and adjust the difficulty of exercises accordingly.

$$S_t = \beta_0 + \beta_1 t + \epsilon_t \tag{13.5}$$

Here, β_0 is a regression coefficient: The intercept of the regression line, representing the expected session score when time (t) is zero.
β_1 is a regression coefficient: The slope of the regression line, indicating how the session score changes with each additional unit of time.
ϵ_t is the error term accounting for the difference between the predicted session score and the actual session score.

The system calculates the regression parameters using historical data to predict future performance, which helps in customizing the rehabilitation plan to suit the patient's improving skill level.

Algorithm 3: Dynamic Difficulty Adjustment
This algorithm adjusts the complexity of the exercises dynamically based on the patient's current performance, employing a feedback loop to ensure optimal challenge levels.

Adjustment Rule:

$$D_{new} = D_{current} + \Delta \tag{13.6}$$

Here,

$D_{current}$: Current difficulty level
D_{new}: The new difficulty level to be set based on the patient's performance
Δ: Change in difficulty level based on performance.

If the accuracy score is above 0.8, increase the difficulty by 1 (Δ=+1).
If the accuracy score is below 0.5, decrease the difficulty by 1 (Δ=−1).
Otherwise, keep the difficulty the same (Δ=0).

This dynamic adjustment ensures that the exercises are appropriately challenging, helping to maximize therapeutic benefits and keep the patient engaged. These mathematical tools and algorithms are integral to monitoring and adapting the ARPT system. They ensure that each session is tailored to the patient's current capabilities and progress, facilitating effective and responsive rehabilitation. The system's ability to adjust difficulty levels in real time based on precise mathematical assessments helps optimize recovery paths and enhances engagement by keeping the exercises within an achievable challenge range.

13.4.3 SAR in the Metaverse

A detailed methodology that outlines hardware and software components, how this system operates, focusing on autonomous decision-making, interaction dynamics, and technical integration for a comprehensive rehabilitation program.

a. **Hardware and Software Components**
b. **Detailed Methodology of SAR in Metaverse Environment**

- **System Setup and Initial Configuration:**

Before beginning therapy, an initial assessment is conducted to understand the patient's specific rehabilitation needs that is shown in Figure 13.3. This includes physical capabilities, rehabilitation goals, and any specific preferences or requirements. The SAR and the VR or AR environments are calibrated to ensure seamless interaction. This includes setting up the visual and sensory outputs of the VR/AR systems and ensuring the robot can navigate and interact within these settings.

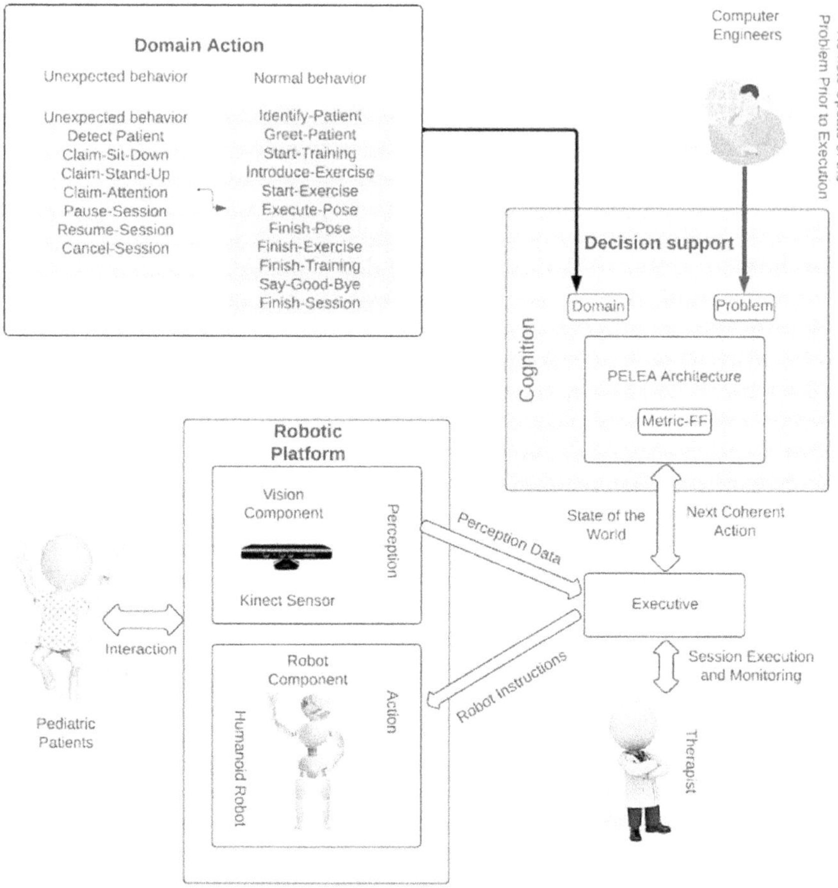

Figure 13.3 An architectural overview illustrates the data flow among all system components and their interactions for SAR in the metaverse environment [09].

- **Customization of Rehabilitation Sessions:**

Therapists interact with a graphical interface to select and customize the rehabilitation exercises based on the patient's profile. This might include choosing certain games or tasks that are best suited to the patient's therapeutic needs. These selections are sent to the engineer or the technical team, which configures the platform remotely, updating and preparing the system with the chosen settings and exercises.

- **Autonomous Operation and Real-Time Interaction:**

Using automated task planning based on artificial intelligence, the system operates autonomously during sessions. It uses a declarative predicate-based representation (possibly in Planning Domain Definition Language (PDDL)) to outline rehabilitation sessions in terms of actions, preconditions, and effects. Robots interact with patients by providing instructions, support, and feedback. This interaction is enhanced by the robot's ability to interpret and respond to patient actions and verbal cues in real-time.

- **Execution of Rehabilitation Activities:**

The decision-support component carries out task planning using a subarchitecture designed for real-time applications (e.g., PELEA). This involves generating a coherent sequence of actions based on the planned therapy session. As the patient engages with the therapy activities, the robot monitors their performance using sensors and cameras. Feedback is provided immediately to encourage correct movement and posture.

- **Adaptive Response and Replanning:**

The system is designed to handle nondeterministic elements or unexpected patient responses. If an activity does not proceed as planned, the decision-support system detects inconsistencies and initiates a replanning process to adjust the activities accordingly. Throughout the session, the system adapts to the patient's responses. For example, if a patient struggles with a specific pose or movement, the robot can modify the task difficulty or provide additional guidance.

- **Evaluation and Progress Tracking:**

Post-session, all interaction and performance data are collected and analyzed to evaluate the patient's progress and the efficacy of the therapy. Based on this analysis, therapists may adjust future sessions to better align with the patient's progress and rehabilitation goals.

- **Patient and Therapist Feedback Loop:**

Feedback from both the patient and the therapist is collected to assess the satisfaction with the session and any areas for improvement. Feedback can lead to further adjustments and updates to the system, ensuring that the therapy remains effective and responsive to patient needs.

13.5 ADDRESSING CHALLENGES

Addressing the challenges of integrating the metaverse into rehabilitation involves tackling a variety of technological, ethical, and practical issues.

Table 13.4 Hardware and software specifications for socially assistive robots in metaverse

Component Type	*Hardware/Software*	*Description*	*Application in System*
Hardware	Head-Mounted Displays (HMDs)	Devices that provide immersive visual and auditory experiences	Used by patients to interact within the VR or AR environments
	Motion Tracking Sensors	Sensors that capture detailed movement data	Allow for precise tracking of patient movements during therapy sessions
	Cameras and Microphones	Capture visual and audio inputs for interaction and monitoring	Enable robots to see, hear, and respond to patient actions and verbal cues
	Socially Assistive Robots (e.g., NAO, Pepper)	Robots equipped with AI for interaction and mobility	Engage directly with patients to guide, motivate, and provide feedback
Software	Virtual/AR Software	Manages the creation and display of virtual environments	Generates immersive therapy environments tailored to patient needs
	AI-driven Planning Software (e.g., PELEA)	Manages task planning and execution within the system	Supports autonomous operation of the robots by planning, monitoring, and adjusting therapy sessions
	Custom Therapy Applications	Designed specifically for rehabilitation tasks and games	Facilitate therapeutic exercises and activities that are interactive and engaging
	Communication Interfaces	Software that facilitates integration and communication between different system components	Ensures seamless interaction between the hardware components and the software layers

13.5.1 High Cost

Despite the significant advancements and benefits brought by metaverse technologies in rehabilitation, several challenges remain. High costs associated with advanced technology deployment, technical complexity, and the need for specialized training for healthcare providers are substantial barriers [15, 16]. These challenges can limit the widespread adoption of such technologies, particularly in resource-limited settings.

13.5.2 Long-term Clinical Trial

Moreover, the clinical validation of these technologies often requires comprehensive, long-term studies to establish their efficacy and safety across diverse patient populations. As noted by Matt C. Howard et al. [15], there is a crucial need for ongoing research to better understand the long-term impacts and potential unintended consequences of using VR, AR, and other metaverse technologies in rehabilitation.

13.5.3 Integration with Traditional Rehabilitation

While metaverse technologies offer exciting prospects, their integration with traditional rehabilitation methods needs careful consideration. The combination of high-tech solutions with conventional therapies could cater to a broader spectrum of rehabilitation needs, ensuring that all patients benefit from the best possible approaches tailored to their specific circumstances.

13.5.4 Ethical and Privacy Concerns

As with any technology that handles sensitive health data, there are ethical and privacy concerns associated with metaverse technologies. Ensuring the security of patient data and addressing potential ethical issues related to virtual treatments are critical. It's essential for developers and healthcare providers to adhere to stringent data protection standards and ethical guidelines to maintain trust and protect patient rights.

13.6 FUTURE DIRECTIONS

As metaverse technologies continue to evolve, the landscape of rehabilitation is poised for transformative changes. Future developments in this field must focus on several key areas to maximize the potential of these technologies and address the challenges that currently limit their widespread adoption.

13.6.1 Technological Enhancements

Future technological enhancements should prioritize user experience and inclusivity. This includes developing more intuitive user interfaces and versatile applications that can accommodate a wider range of disabilities and rehabilitation needs. Advances in AI and machine learning can play a significant role here, particularly in creating adaptive systems that respond dynamically to the user's progress and needs. Integrating AI more deeply with VR, AR, and MR could lead to more personalized therapy sessions that adjust in real-time to optimize therapeutic outcomes based on continuous feedback from the patient's performance.

13.6.2 Cost Reduction

One of the primary barriers to the widespread adoption of metaverse technologies in rehabilitation is cost. Future efforts must focus on reducing the cost of VR and AR hardware, as well as the development costs associated with creating and maintaining high-quality therapeutic content. Economies of scale and more streamlined production processes might help lower these costs. Additionally, open-source software and hardware solutions could be explored to provide low-cost alternatives for budget-constrained healthcare facilities.

13.6.3 Clinical Trials

To validate the efficacy and safety of metaverse technologies in rehabilitation, comprehensive clinical trials are crucial. These studies should not only assess the effectiveness of these technologies in improving patient outcomes but also examine their long-term impacts on physical and mental health. Multicenter trials involving diverse patient populations will be essential to ensure that the findings are generalizable across different demographics and medical conditions.

13.6.4 Interdisciplinary Collaboration

Encouraging closer collaboration among technologists, healthcare professionals, and researchers from various disciplines can accelerate innovation and adoption. This collaboration can help ensure that technological developments are aligned with clinical needs and are based on sound medical research.

13.6.5 Ethical and Regulatory Considerations

As metaverse technologies become more integrated into healthcare, ethical and regulatory frameworks must be developed and refined to manage issues related to patient privacy, data security, and the ethical use of AI and other advanced technologies in clinical settings.

Looking forward, the development of more cost-effective, user-friendly, and robust metaverse technologies will be crucial. Additionally, fostering interdisciplinary collaborations among technologists, clinicians, and researchers can drive innovations that are both technologically advanced and clinically relevant.

Through this discussion, it becomes evident that while metaverse technologies hold transformative potential for rehabilitation, realizing this potential fully requires addressing the existing challenges through innovative solutions, rigorous research, and collaborative efforts across various domains.

13.7 CONCLUSION

The integration of metaverse technologies into rehabilitation represents a significant paradigm shift toward more personalized, engaging, and accessible therapy options. These technologies offer the potential to transform traditional rehabilitation by providing immersive, interactive experiences that can motivate patients and lead to better outcomes. However, the full realization of this potential is contingent upon overcoming several barriers.

Current challenges include the high costs associated with these technologies, the technical complexity of implementing and maintaining advanced systems, and the need for extensive clinical validation to ensure their effectiveness and safety across diverse patient groups. Additionally, ethical considerations related to data privacy and the equitable distribution of these technologies must be addressed to avoid exacerbating existing healthcare disparities.

Looking forward, the focus should be on enhancing the technological capabilities of metaverse systems, reducing costs, and conducting rigorous clinical trials. By addressing these issues, it will be possible to fully leverage the benefits of metaverse technologies in rehabilitation settings. This will not only improve the quality of care available to patients but also expand access to rehabilitation services globally, ultimately contributing to better health outcomes and enhanced quality of life for individuals with disabilities.

The journey ahead is promising, but it requires concerted efforts from all stakeholders involved in healthcare technology, policy making, and clinical practice. Through continued innovation, collaboration, and diligent attention to the ethical dimensions of technology use in healthcare, the future of rehabilitation in the metaverse can be as transformative as it is inclusive.

REFERENCES

1. Gotsis, M., et al. (2012). Mixed reality game prototypes for upper body exercise and rehabilitation. 2012 IEEE Virtual Reality Workshops (VRW), Costa Mesa, CA, 181–182. https://doi.org/10.1109/VR.2012.6180940
2. Høeg, E. R., Bruun-Pedersen, J. R., Cheary, S., et al. (2023). Buddy biking: A user study on social collaboration in a virtual reality exergame for rehabilitation. *Virtual Reality, 27*(2), 245–262. https://doi.org/10.1007/s10055-021-00544-z
3. Elor, A., Teodorescu, M., & Kurniawan, S. (2018). Project Star Catcher: A novel immersive virtual reality experience for upper limb rehabilitation. *ACM Transactions on Accessible Computing, 11*(4), Article 20. https://doi.org/10.1145/3265755
4. Mann, S., Furness, T., Yuan, Y., Iorio, J., & Wang, Z. (2018). All reality: Virtual, augmented, mixed (X), mediated (X,Y), and multimediated reality. *arXiv*. https://arxiv.org/abs/1804.08386
5. Gómez-Portes, C., Carneros-Prado, D., Albusac, J. J., Castro-Sánchez, J., Glez-Morcillo, C., & Vallejo, D. (2021). PhyRe Up! A system based on mixed reality and gamification to provide home rehabilitation for stroke patients. *IEEE Access, 9*, 139122–139137. https://doi.org/10.1109/ACCESS.2021.3118842
6. Yang, Z. Q., Du, D., Wei, X. Y., et al. (2022). Augmented reality for stroke rehabilitation during COVID-19. *Journal of NeuroEngineering and Rehabilitation, 19*, 136. https://doi.org/10.1186/s12984-022-01100-9
7. Lund, E. M., Forber-Pratt, A. J., Wilson, C., & Mona, L. R. (2020). The COVID-19 pandemic, stress, and trauma in the disability community: A call to action. *Rehabilitation Psychology, 65*(4), 313–322. https://doi.org/10.1037/rep0000368
8. Jacob, S., et al. (2021). AI and IoT-enabled smart exoskeleton system for rehabilitation of paralyzed people in connected communities. *IEEE Access, 9*, 80340–80350. https://doi.org/10.1109/ACCESS.2021.3083093
9. Sobrepera, M. J., Lee, V. G., & Johnson, M. J. (2021). The design of Lil'Flo, a socially assistive robot for upper extremity motor assessment and rehabilitation in the community via telepresence. *Journal of Rehabilitation and Assistive Technologies Engineering, 8*, 20556683211001805. https://doi.org/10.1177/20556683211001805
10. Bortone, I., Leonardis, D., Mastronicola, N., Crecchi, A., Bonfiglio, L., Procopio, C., Solazzi, M., & Frisoli, A. (2018). Wearable haptics and immersive virtual reality rehabilitation training in children with neuromotor impairments. *IEEE Transactions on Neural Systems and Rehabilitation Engineering, 26*(7), 1469–1478. https://doi.org/10.1109/TNSRE.2018.2846814
11. Pulido, J. C., et al. (2019). A socially assistive robotic platform for upper-limb rehabilitation: A longitudinal study with pediatric patients. *IEEE Robotics & Automation Magazine, 26*(2), 24–39. https://doi.org/10.1109/MRA.2019.2905231

12. Hudák, M., Sobota, B., Korečko, Š., & Sivý, M. (2020). Fully immersive web-based collaborative virtual environment for upper limb rehabilitation purposes. 2020 18th International Conference on Emerging eLearning Technologies and Applications (ICETA), Košice, Slovenia, 194–199. https://doi.org/10.1109/ICETA51985.2020.9379239
13. Inamura, T., & Mizuchi, Y. (2020). SIGVerse: A cloud-based VR platform for research on social and embodied human-robot interaction. *arXiv.* https://arxiv.org/abs/2005.00825
14. Maier, M., Rubio Ballester, B., Duarte, E., Duff, A., & Verschure, P. F. M. J. (2014). Social integration of stroke patients through the multiplayer rehabilitation gaming system. In S. Göbel & J. Wiemeyer (Eds.), *Games for Training, Education, Health and Sports: GameDays 2014* (pp. 147–164). Springer. https://doi.org/10.1007/978-3-319-05972-3_12
15. Howard, M. C., & Davis, M. M. (2022). A meta-analysis and systematic literature review of mixed reality rehabilitation programs: Investigating design characteristics of augmented reality and augmented virtuality. *Computers in Human Behavior, 130*, 107197. https://doi.org/10.1016/j.chb.2022.107197
16. Specian, A., Mead, R., Kim, S., Matarić, M., & Yim, M. (2022). Quori: A community-informed design of a socially interactive humanoid robot. *IEEE Transactions on Robotics, 38*(3), 1755–1772. https://doi.org/10.1109/TRO.2021.3111718
17. Sutherland, I. E. (1968). A head-mounted three-dimensional display. Fall Joint Computer Conference.
18. Rothbaum, B. O., Hodges, L., Smith, S., Lee, J. H., & Price, L. (1995). A controlled study of virtual reality exposure therapy for the fear of flying. *Journal of Consulting and Clinical Psychology, 63*(2), 202–211.
19. Cruz-Neira, C., Sandin, D. J., & DeFanti, T. A. (1993). Surround-screen projection-based virtual reality: The design and implementation of the CAVE. Proceedings of the 20th Annual Conference on Computer Graphics and Interactive Techniques, 135–142.
20. Kato, H., & Billinghurst, M. (1999). Marker tracking and HMD calibration for a video-based augmented reality conferencing system. Proceedings of the 2nd IEEE and ACM International Workshop on Augmented Reality, 85–94.
21. Mataric, M. J., Tapus, A., Winstein, C. J., & Eriksson, J. (2011). Socially assistive robotics for post-stroke rehabilitation. *Journal of NeuroEngineering and Rehabilitation, 8*(24).
22. Lange, B., Flynn, S., & Rizzo, A. (2012). Initial usability assessment of off-the-shelf video game consoles for clinical game-based motor rehabilitation. *Physical Therapy Reviews, 17*(2), 98–109.
23. Patel, S., Park, H., Bonato, P., Chan, L., & Rodgers, M. (2014). A review of wearable sensors and systems with application in rehabilitation. *Journal of NeuroEngineering and Rehabilitation, 11*, 70.
24. Loch, F., & Naeemabadi, M. R. (2016). Mixed reality applications in industry: Challenges and research areas. *Virtual Reality, 20*(3), 183–200.

25. Zhao, Y., Nonnekes, J., Storcken, E. J., Janssen, S., van Wegen, E. E., Bloem, B. R., Dorresteijn, T. A., & van Vugt, J. P. (2018). Feasibility of external rhythmic cueing with the Google Glass for improving gait in people with Parkinson's disease. *Journal of NeuroEngineering and Rehabilitation, 15*, 74.
26. Lee, H. M., Chung, H., Ko, H., Lee, J. K., Lee, W. H., Lee, G., Shin, J. H., & Lee, J. (2020). Feasibility of virtual reality rehabilitation for mobility and cognition in elderly patients with neurological diseases. *Journal of Clinical Medicine, 9*(2), 586.
27. Garcia, A. D., Saenz, N., & Gaviria, M. (2022). Effective virtual reality interventions for neuropathic pain management: A systematic review. *Pain Research and Management, 2022*, 2348197. https://doi.org/10.1155/2022/2348197

Chapter 14

Virtual Reality Medical Technology in the Metaverse

A Digital Twin Approach and Cybersecurity Considerations

Jingyang Cui and Shengting Cao

14.1 INTRODUCTION

As medical technology continues to advance, virtual reality (VR) [1] and digital twin (DT) [2] have shown great potential as innovative tools in the medical field. VR technology enhances diagnosis and treatment by creating an immersive environment that allows patients and medical professionals to interact in a virtual space. DT technology provides accurate data support for medical processes through real-time modeling and simulation of physical entities.

In medical applications, DT technology can reflect a patient's physiological state and treatment progress in real time, enabling medical personnel to better understand a patient's needs. This data can be visualized with VR technology to help doctors make more intuitive assessments and treatment plans in a virtual environment. In recent years, many studies and practice cases have shown that this combination can not only improve medical efficiency but also improve the patient experience.

However, despite the promising application of VR and DTs in medical technology, there are also many challenges in the actual implementation process. These challenges include the integration of technologies, real-time processing of data, optimization of user experience, and security issues. Therefore, it is of great theoretical significance and practical value to study the framework of VR medical technology based on DTs and its application in different medical scenarios.

This chapter will discuss the basic principle, application status, and future development direction of VR medical technology based on DT. Through the review and analysis of relevant literature, we will establish a clear research framework and put forward the corresponding methodology, in order to provide useful reference for related research.

The rest of this chapter is organized as follows. In the "Related works" section, some related works are introduced. The "Methodology" section mainly introduces how to build a DT medical system. "Cybersecurity Problems at VRMT" includes network security issues faced by virtual medical technology. The "Future work" section discusses the current problems

DOI: 10.1201/9781032663609-17

and possible future research directions. The "Conclusion" section briefly summarizes the full text and makes further prospects for future work.

14.2 RELATED WORKS

In recent years, the application of VR and DT technologies in the medical field has garnered increasing attention from both academia and industry. Relevant studies have primarily focused on three key areas: the use of VR in medical simulation and training, the role of DT in personalized healthcare and diagnostics, and the potential of combining both technologies for medical visualization and telemedicine [3].

14.2.1 The Application of VR in Healthcare

VR technology has been widely applied in areas such as medical simulation, surgical training, and rehabilitation therapy. For example, McKnight et al. [4] integrated VR with a surgical simulation system, enabling virtual training of surgeons' skills and significantly reducing the error rate during surgeries. Compared to traditional training methods, VR offers a more immersive environment, allowing physicians to practice high-risk procedures in a low-risk setting. Additionally, VR has shown great success in psychotherapy and rehabilitation, such as exposure therapy for treating phobias and the design of virtual rehabilitation environments [5].

14.2.2 The Role of DT in Healthcare

DT technology offers new possibilities for personalized medicine, particularly in real-time patient monitoring and treatment optimization. Sack et al. [6] developed a heart simulation model based on DT that can simulate a patient's blood flow and electrical signals in real time, providing instant feedback to doctors and aiding in the creation of personalized treatment plans. DT has also played a significant role in disease prediction and early intervention. For example, Drummond et al. [7] built a data-driven DT model to monitor and manage the health of children with chronic diseases, offering strong support for precision medicine.

14.2.3 The Prospects of Combining VR and DT

The integration of VR and DT technologies makes complex medical data visualization possible and presents new solutions for telemedicine. Liu et al. [8] demonstrated that by combining a patient's DT with a VR system; doctors can remotely monitor the patient's health and simulate and optimize treatments in a virtual environment. This technological advancement is particularly valuable for remote areas and underserved healthcare environments, significantly improving the accuracy and efficiency of telemedicine. Kim et al. [9] developed a remote consultation system that combines DT with VR, allowing healthcare professionals to enhance diagnostic accuracy using digital models of patients and auxiliary examination tools.

Despite the progress in this area, there are still significant challenges in the widespread adoption of VR and DT technologies in healthcare, including issues related to data processing capabilities, computational resource demands, and user interaction experience. Future research should focus on improving the integration of these technologies and exploring their potential in diverse healthcare scenarios.

14.3 METHODOLOGY

In this study, we propose a DT-based Virtual Reality Medical Technology (DTVRMT) designed to integrate DT models with VR environments to enable real-time monitoring, assessment, and intervention of patient health conditions. The primary focus is on rehabilitation training for individuals with disabilities or stroke patients under telemedicine conditions. The system uses cameras to perform 3D modeling of the patient's rehabilitation process on a treadmill (in Figure 14.1) and implements it into a DT platform, providing remote medical support. The process diagram is shown in Figure 14.2. The system's workflow comprises several key steps: system design, data collection, DT modeling, VR environment construction, and system supervision.

Patients undergo rehabilitation training locally and video recordings are made by cameras, which are processed by deep learning models, resulting in 3D models and uploaded to the metaverse. The telemedicine process involves downloading 3D models from the metaverse and rendering them, which are then presented in VR headsets.

14.3.1 System Design

The system is designed with a modular approach, consisting of three main modules: data collection, DT, and VR. There are seven processing steps. A detailed representation of these steps is shown in Figure 14.3.

The data collection module is responsible for capturing video footage of the patient's rehabilitation process on the treadmill and acquiring real-time data through sensors and medical devices. The DT module uses this data to create a personalized DT model of the patient, simulating their physiological condition, movement, and health changes. Finally, the VR module visualizes the DT model, allowing doctors to perform health assessments and interventions within a virtual environment.

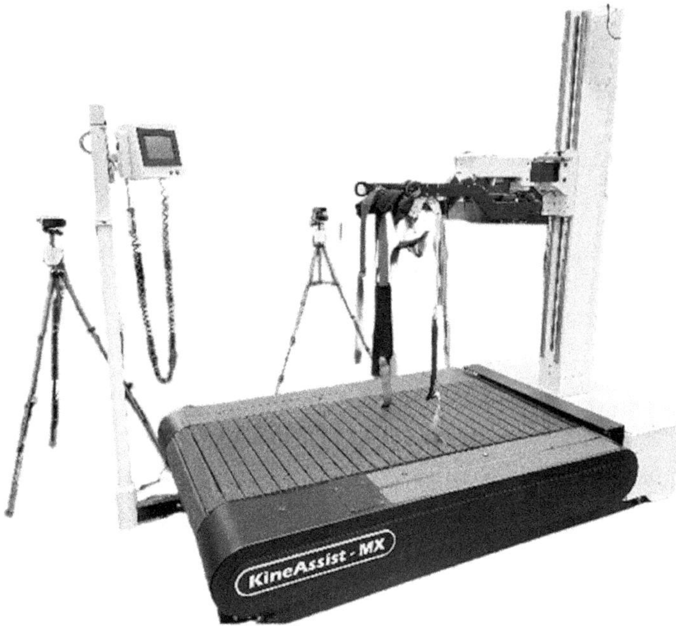

Figure 14.1 Smart treadmill.

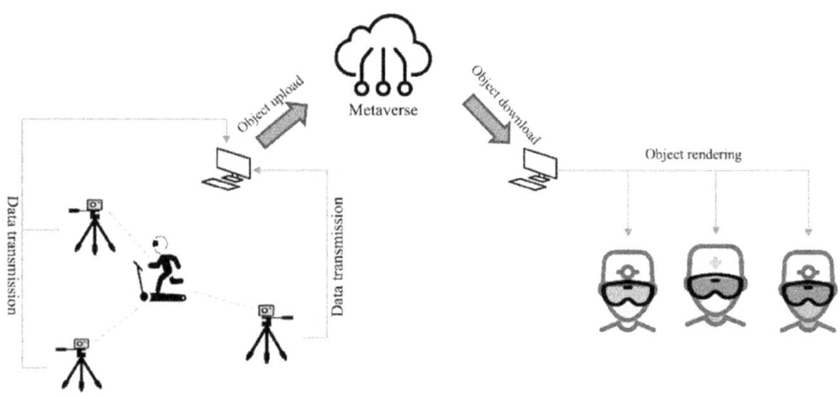

Figure 14.2 Telehealth process based on metaverse.

The treatment process in a telemedicine system generally consists of seven steps.

First, the system environment and equipment are configured. For video transmission from cameras, both wired and wireless options are feasible. Notably, in our experiments, we found that wired transmission requires high-quality cables; some cables experience significant signal degradation over distances exceeding two meters, causing substantial image quality issues or even transmission failure. While wireless transmission is more challenging to implement, it provides better data quality. Next, the target subject is scanned. Using cameras with radar functionality, we collect and organize necessary video information, which allows for real-time distance detection and immediate feedback.

After this, deep learning methods are applied for modeling. Specifically, we use Sam2 [10] to segment the scanned subject semantically, identify dynamic elements, remove the background, and construct a 3D model.

The fourth step is model rendering, where we process the model and images within a Unity-based rendering framework using the Instant [11] algorithm.

For data transmission, we currently conduct experiments within a limited area, using wired transmission. Expanding the range and distance for larger-scale applications may introduce time synchronization issues, which we aim to address in future research.

Following data transmission, we deploy the treatment scenario in VR headsets, enabling doctors to observe patients' movements and trajectories comprehensively.

The final step involves third-party review and learning from the treatment process. We consider two main contexts here: sharing treatment experiences with others and enabling oversight for regulatory purposes.

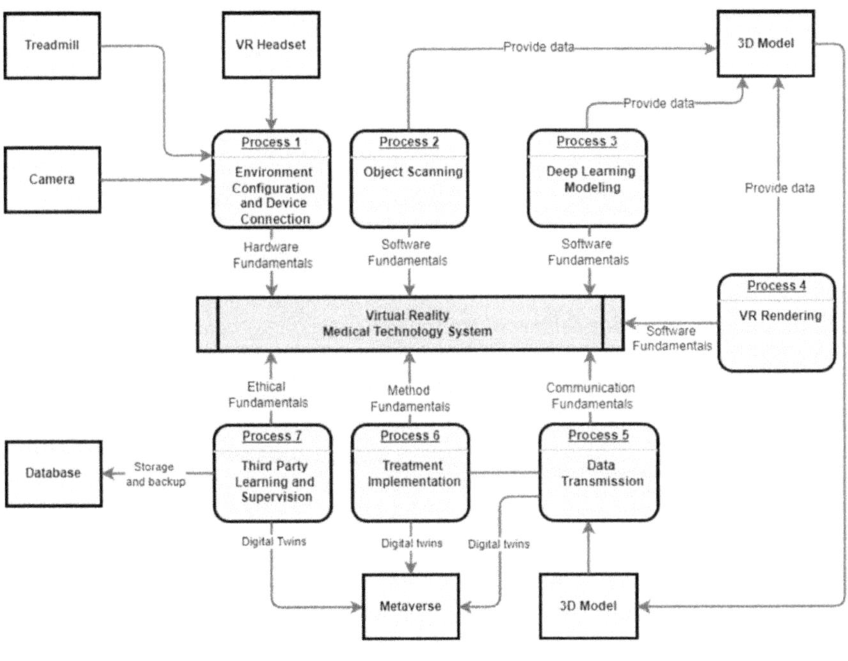

Figure 14.3 System process.

14.3.2 Data Collection

To ensure the accuracy of the DT model, various sensors such as heart rate monitors, accelerometers, and physiological monitoring devices are employed for data collection. The treadmill used in rehabilitation also generates a large amount of sensor data. These sensors are connected to the system via a network, enabling real-time transmission of the patient's health data. The data generated by the collection system is stored in a database for further analysis and processing. For the camera, we used Azure Kinect DK, as shown in Figure 14.4. Azure Kinect DK is an RGB+ depth camera suite from Microsoft for spatial awareness and human-computer interaction. It integrates a variety of sensors, including a depth camera, an RGB image camera, a microphone array, and an accelerometer.

14.3.3 DT Modeling

The DT modeling process involves data preprocessing, modeling, and validation. First, the collected physiological data is preprocessed, including noise reduction, standardization, and data completion. Then, based on the processed data, machine learning and data mining techniques are used to build the DT model. These models can reflect the patient's health status in real time and predict future health changes. Finally, the accuracy and reliability of the DT model are validated by comparing it with actual clinical data. Figure 14.5 is the 3D model created by us after scanning, and the image quality has been enhanced.

14.3.4 VR Environment Construction

The VR environment is built using the Unity engine and incorporates HoloLens 2 for an immersive experience. In this environment, doctors can interact with the patient's DT, performing health assessments and treatment simulations. A series of virtual scenes were designed to allow doctors to observe the patient's movements in detail, making it easier to evaluate the treatment effects under different scenarios. Based on these results we built a prototype telemedicine system using Unity, and in this system, doctors can watch a patient's recovery process through Headset. Figure 14.6 shows the doctor's first perspective. The doctor can enlarge and rotate the 3D model by hand.

14.3.5 System Validation

To assess the system's effectiveness and usability, we conducted a multistage testing and validation process. First, simulated tests were performed in a laboratory environment to verify the accuracy and stability of data collection, modeling, and visualization. Next, we collaborated with clinical

Figure 14.4 Azure Kinect DK.

doctors to conduct field tests with real patients, gathering user feedback to optimize the system's design and functionality. Finally, comparative experiments were performed to evaluate the differences in treatment outcomes and patient satisfaction between VRMT and traditional medical methods.

14.3.6 The Observation of Telemedicine

We designed a module to allow third parties to observe the treatment process. In Figure 14.7, two objects represent the doctor and the patient, rendered in Unity. By entering these two cubes, we can observe the first-person rendering perspective of each. In future stages, we plan to establish a third-person full-body view of both the doctor and patient to provide a more intuitive observation of the treatment process.

Incorporating VR and DT technology enables a collaborative, multidisciplinary approach to treatment where experts can conduct consultations, review cases, and provide remote guidance for complex procedures. This setup is particularly beneficial for training new medical professionals, offering them firsthand experience and insights into best practices from senior specialists. By leveraging virtual platforms, remote treatments can align with national policies and regulatory standards, helping ensure patient safety and data protection. Additionally, robust cybersecurity measures are integrated to mitigate potential threats, minimizing the risks of

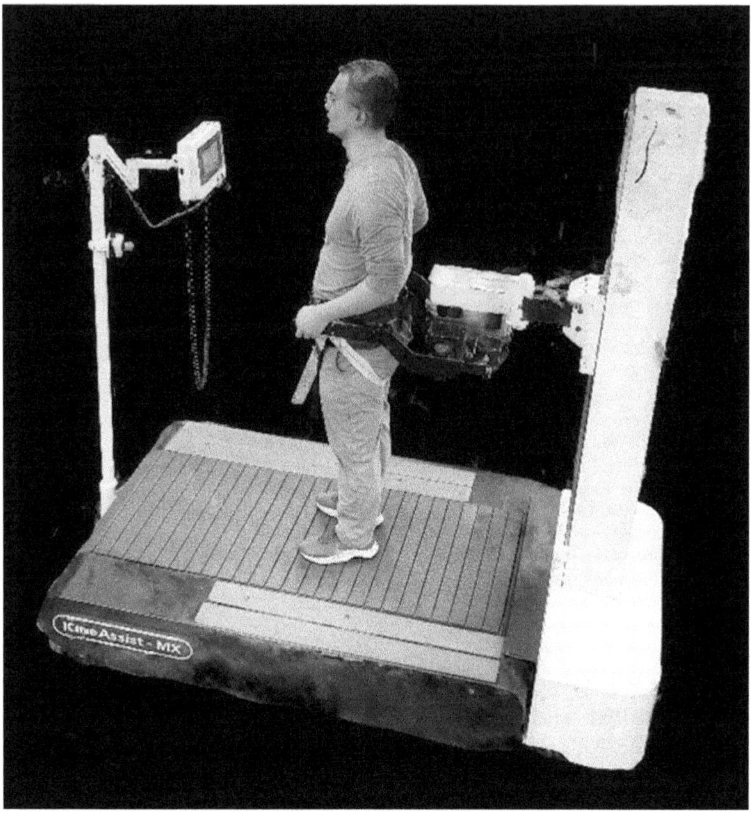

Figure 14.5 3D model result.

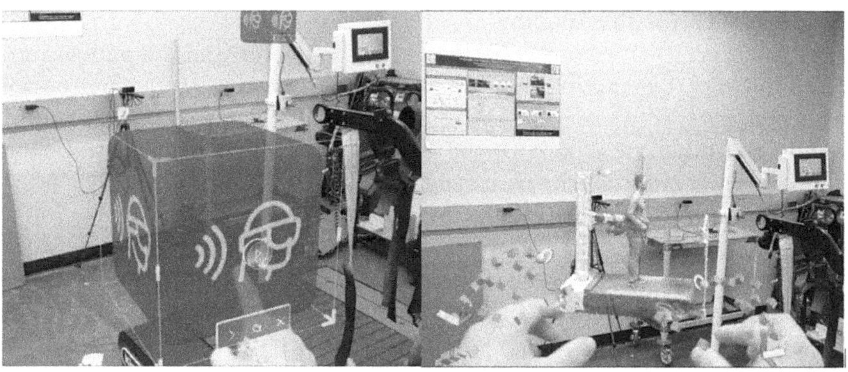

Figure 14.6 3D model rendering process.

unauthorized access and cybercrimes, and maintaining a secure and compliant treatment environment.

14.4 CYBERSECURITY PROBLEMS AT VRMT

VRMT in the Metaverse, particularly through devices such as HoloLens, Kinect, and Meta Quest, offers significant potential for remote diagnosis and assessment for stroke rehabilitation patients. This application, based on the metaverse and DT technology, not only shows great promise in the medical field but also introduces several cybersecurity risks, primarily in the following areas:

14.4.1 Data Transmission and Privacy Leakage

During the transfer of patient data, such as 3D holographic avatars and motion data, to the doctor's device, large amounts of Personal Health Information (PHI) are involved. If these transmissions are not adequately encrypted, sensitive information may be intercepted or stolen, violating patient privacy [12, 13–14]. This concern is especially critical in DT and metaverse environments, where real-time, detailed physiological data is processed.

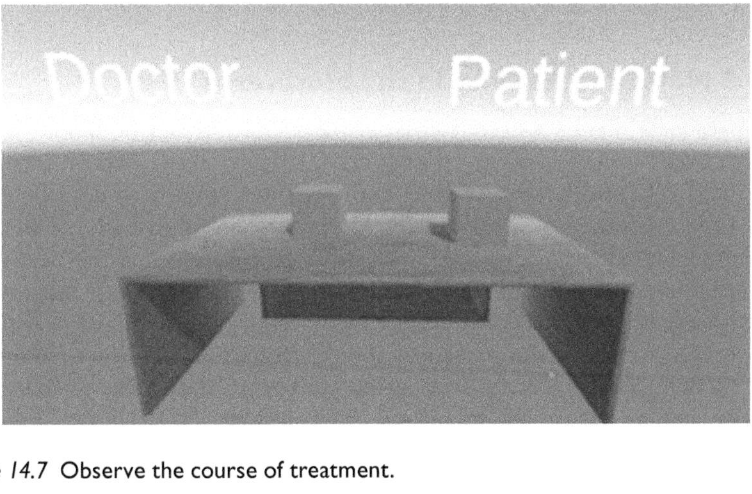

Figure 14.7 Observe the course of treatment.

14.4.2 Device and Network Security

The system utilizes multiple devices, such as HoloLens 2, Kinect cameras, and connected treadmills, all of which are networked. If any device or network node has a vulnerability, hackers could exploit these weaknesses to breach the system, alter data, or disrupt the communication between patients and doctors [15, 16]. For example, an attack on the HoloLens devices used by either the doctor or the patient could allow the attacker to manipulate the displayed content, disrupt the treatment process, or even affect the patient's rehabilitation outcomes.

14.4.3 Authentication and Access Control

In this system, both doctors and patients interact through virtual avatars, which introduces the risk of identity theft [17]. Without strict authentication mechanisms, an attacker could impersonate either the doctor or the patient, gaining unauthorized access to sensitive information or interfering with the rehabilitation process [18]. It is essential to ensure that only authorized doctors and patients have access to the system and its devices.

14.4.4 DT System Integrity and Security

The system creates a real-time DT (a digital replica) of the patient based on collected data. Ensuring the authenticity and integrity of this data is crucial[19]. If the DT is tampered with, doctors may make incorrect diagnoses or assessments based on false information. Therefore, it is vital to prevent hackers from altering or injecting false data into the DT system.

14.4.5 Malicious Activity in the Metaverse

The metaverse, being a shared virtual space, may attract malicious actors. If the system is integrated into a metaverse environment, it could introduce additional security threats, such as malicious disruptions or data hijacking within the virtual environment [20]. These actions could ruin the patient's rehabilitation experience or lead to the corruption or loss of medical data.

To address these cybersecurity issues, healthcare institutions must implement robust security measures when introducing similar mixed reality and metaverse systems. These measures should include, but are not limited to, data encryption, device authentication, intrusion detection systems, and security regulations for the metaverse environment. This will not only protect patient privacy but also ensure that doctors can make accurate medical decisions based on reliable data.

14.4.6 Cybersecurity Considerations Conclusion

In the application of VRMT, cybersecurity is crucial and primarily involves the following aspects:

i. **Data Privacy:** Patients' physiological and medical data are at risk of exposure during transmission and storage. Unauthorized access to this sensitive information could significantly impact patients' privacy and sense of security.
ii. **Device Security:** VRMT systems rely on the coordinated operation of multiple devices, including sensors, cameras, and computing devices. Without adequate protection, these devices may be vulnerable to cyberattacks, potentially leading to system crashes or data loss.
iii. **Risk of Network Attacks:** VRMT requires remote interaction over the internet, which exposes it to various types of network attacks, such as Denial of Service (DoS) and Man-in-the-Middle (MITM) attacks. These attacks could not only disrupt medical services but also result in data tampering.

To address the cybersecurity challenges above, the following solutions are recommended:

i. **Encryption Technology:** Encrypt data during transmission to ensure that intercepted information cannot be interpreted by unauthorized parties.
ii. **Access Control:** Implement strict user authentication mechanisms to ensure that only authorized personnel can access sensitive data and system functionalities.
iii. **Monitoring and Response Mechanism:** Establish a real-time monitoring system to promptly detect and respond to potential security threats, ensuring the system operates smoothly.

14.5 FUTURE WORK

Although this study has made significant progress in the application of VRMT based on DT technology, there remain several areas for improvement and expansion. Future research could focus on the following aspects:

14.5.1 Enhancing Model Accuracy

While our DT model has shown promising predictive accuracy, there is still room for improvement. Future work will focus on the following:

- **Multi-source Data Integration:** Integrate data from various devices (e.g., wearables, medical imaging devices) to enhance the model's comprehensiveness and accuracy.
- **Algorithm Optimization:** Explore advanced machine learning and deep learning algorithms (e.g., Graph Convolutional Networks) to improve the model's responsiveness to complex changes in health conditions.

14.5.2 Expanding the Scope of Application

Current research primarily focuses on stroke rehabilitation, and future efforts will explore VRMT's application in other healthcare fields, including but not limited to:

- **Chronic Disease Management:** For conditions such as diabetes and heart disease, enabling real-time monitoring and virtual intervention to improve patients' self-management abilities.
- **Mental Health Therapy:** Using VR technology to intervene in and treat mental health issues such as anxiety and depression.

14.5.3 Improving User Experience

User experience in VR environments is crucial to the system's success. Future research will consider the following:

- **User Interface Optimization:** Collect user feedback through testing to design a more user-friendly interface, enhancing interaction for both clinicians and patients.
- **Enhanced Immersion:** Improve the realism of the virtual environment with higher-quality visuals and audio, making it easier for patients to adapt to and accept treatment.

14.5.4 Data Security and Privacy Protection

With the increase in digital health data, ensuring patient data security and privacy will be critical. Future work will emphasize:

- **Data Encryption:** Ensuring security during data transmission and storage to prevent data breaches.
- **Privacy Policy Development:** Formulate and refine policies for data usage and storage in line with relevant laws and regulations to enhance patient trust in the system.

14.5.5 Clinical Trials and Commercialization

To validate VRMT's effectiveness on a larger scale, future research will involve larger clinical trials. Additionally, pathways to commercialize the technology will be explored, including:

- **Partnerships with Healthcare Institutions:** Establish collaborations with hospitals, clinics, and other healthcare institutions to better promote and apply the technology.
- **Market Demand Analysis:** Conduct a thorough analysis of market needs to assess VRMT's commercial potential and development prospects.

14.6 CONCLUSION

This study examines the application and development of VRMT based on DT technology, demonstrating its potential and advantages in stroke rehabilitation. By constructing a DT model, we are able to monitor and assess patients' physiological conditions in real time and conduct remote rehabilitation therapy via VR. This approach not only enhances treatment efficiency but also reduces travel costs for patients, providing an innovative solution for accessible healthcare services.

Through a review of related work and analysis of existing technologies, we confirm the feasibility of VRMT in practical applications. However, current research faces several challenges, including model accuracy, user experience, and data security. Future efforts will address these issues to further improve the effectiveness of this technology.

As medical technology advances, DT-based VRMT is expected to play an essential role in additional healthcare fields. Through continuous research and optimization, we hope VRMT can offer more convenient and effective healthcare services to a broader range of patients, ultimately improving their quality of life and health outcomes.

In summary, VRMT not only promotes the application of VR and DT technologies in healthcare but also offers new directions for future medical innovations and reform.

REFERENCES

1. M. Javaid and A. Haleem, "Virtual reality applications toward medical field," *Clin. Epidemiol. Glob. Health*, vol. 8, no. 2, pp. 600–605, Jun. 2020, doi: 10.1016/j.cegh.2019.12.010.

2. T. Sun, X. He, and Z. Li, "Digital twin in healthcare: Recent updates and challenges," *Digit. Health*, vol. 9, p. 20552076221149651, Jan. 2023, doi: 10.1177/20552076221149651.
3. M. W. Zackoff *et al.*, "Immersive virtual reality onboarding using a digital twin for a new clinical space expansion: A novel approach to large-scale training for health care providers," *J. Pediatr.*, vol. 252, pp. 7–10.e3, Jan. 2023, doi: 10.1016/j.jpeds.2022.07.031.
4. R. R. McKnight, C. A. Pean, J. S. Buck, J. S. Hwang, J. R. Hsu, and S. N. Pierrie, "Virtual reality and augmented reality—translating surgical training into surgical technique," *Curr. Rev. Musculoskelet. Med.*, vol. 13, no. 6, pp. 663–674, Dec. 2020, doi: 10.1007/s12178-020-09667-3.
5. P. Lindner *et al.*, "Experiences of gamified and automated virtual reality exposure therapy for spider phobia: Qualitative study," *JMIR Serious Games*, vol. 8, no. 2, p. e17807, Apr. 2020, doi: 10.2196/17807.
6. K. L. Sack, J. J. Blauer, M. P. Campbell, and D. J. Swenson, "Creating a digital twin to investigate AV block: In-sights from a validated electromechanical full-heart model," in *2020 Computing in Cardiology*, Sep. 2020, pp. 1–4. doi: 10.22489/CinC.2020.238.
7. D. Drummond and A. Coulet, "Technical, ethical, legal, and societal challenges with digital twin systems for the management of chronic diseases in children and young people," *J. Med. Internet Res.*, vol. 24, no. 10, p. e39698, Oct. 2022, doi: 10.2196/39698.
8. Y. Liu *et al.*, "A novel cloud-based framework for the elderly healthcare services using digital twin," *IEEE Access*, vol. 7, pp. 49088–49101, 2019, doi: 10.1109/ACCESS.2019.2909828.
9. T. Kim, H.-S. Kwon, K. Cho, and W. Woo, "Holistic patient assessment system using digital twin for XR medical teleconsultation," in *Proceedings of the Augmented Humans International Conference 2024*, in AHs '24. Association for Computing Machinery New York, NY, May 2024, pp. 72–78. doi: 10.1145/3652920.3652943.
10. N. Ravi *et al.*, "SAM 2: Segment anything in images and videos," Aug. 01, 2024, *arXiv*: arXiv:2408.00714. doi: 10.48550/arXiv.2408.00714.
11. T. Müller, A. Evans, C. Schied, and A. Keller, "Instant neural graphics primitives with a multiresolution hash encoding," *ACM Trans Graph*, vol. 41, no. 4, p. 102:1-102:15, Jul. 2022, doi: 10.1145/3528223.3530127.
12. A. Altameem, V. Kovtun, M. Al-Ma'aitah, T. Altameem, F. H, and A. E. Youssef, "Patient's data privacy protection in medical healthcare transmission services using back propagation learning," *Comput. Electr. Eng.*, vol. 102, p. 108087, Sep. 2022, doi: 10.1016/j.compeleceng.2022.108087.
13. V. Salunkhe, A. Tangudu, C. Mokkapati, P. (Dr) P. Goel, and A. Aggarwal, "Advanced encryption techniques in healthcare IoT: Securing patient data in connected medical devices," *Mod. Dyn. Math. Progress.*, vol. 1, no. 2, Art. no. 2, Aug. 2024, doi: 10.36676/mdmp.v1.i2.22.
14. Z. K. Mohammed *et al.*, "A metaverse framework for IoT-based remote patient monitoring and virtual consultations using AES-256 encryption," *Appl. Soft Comput.*, vol. 158, p. 111588, Jun. 2024, doi: 10.1016/j.asoc.2024.111588.

15. J. Shang and J. Wu, "Secure voice input on augmented reality headsets," *IEEE Trans. Mob. Comput.*, vol. 21, no. 4, pp. 1420–1433, Apr. 2022, doi: 10.1109/TMC.2020.3020470.
16. J. Shang, "Security and privacy of augmented reality systems," in *Network Security Empowered by Artificial Intelligence*, Y. Chen, J. Wu, P. Yu, and X. Wang, Eds. Springer Nature Switzerland, Cham, 2024, pp. 305–330. doi: 10.1007/978-3-031-53510-9_11.
17. D. Dharminder, U. Kumar, and P. Gupta, "A construction of a conformal Chebyshev chaotic map based authentication protocol for healthcare telemedicine services," *Complex Intell. Syst.*, vol. 7, no. 5, pp. 2531–2542, Oct. 2021, doi: 10.1007/s40747-021-00441-7.
18. T.-W. Lin, C.-L. Hsu, T.-V. Le, C.-F. Lu, and B.-Y. Huang, "A smartcard-based user-controlled single sign-on for privacy preservation in 5G-IoT telemedicine systems," *Sensors*, vol. 21, no. 8, Art. no. 8, Jan. 2021, doi: 10.3390/s21082880.
19. C. Alcaraz and J. Lopez, "Digital twin: A comprehensive survey of security threats," *IEEE Commun. Surv. Tutor.*, vol. 24, no. 3, pp. 1475–1503, 2022, doi: 10.1109/COMST.2022.3171465.
20. A. Chaudhari *et al.*, "Cyber security challenges in social meta-verse and mitigation techniques," in *2024 MIT Art, Design and Technology School of Computing International Conference (MITADTSoCiCon)*, Apr. 2024, pp. 1–7. doi: 10.1109/MITADTSoCiCon60330.2024.10575295.

Index

For Product Safety Concerns and Information please contact our EU
representative GPSR@taylorandfrancis.com
Taylor & Francis Verlag GmbH, Kaufingerstraße 24, 80331 München, Germany

www.ingramcontent.com/pod-product-compliance
Ingram Content Group UK Ltd.
Pitfield, Milton Keynes, MK11 3LW, UK
UKHW022220190926
475702UK00003B/470
* 9 7 8 1 0 3 2 6 6 3 5 5 5 *